SALTERS ADVANCED CHEMISTRY

Chemical **AS**
Storylines

D0243775

Central Team

Derek Denby

Chris Otter

Kay Stephenson

www.heinemann.co.uk

✓ Free online support
✓ Useful weblinks
✓ 24 hour online ordering

01865 888080

In Exclusive Partnership

Heinemann is an imprint of Pearson Education Limited,
a company incorporated in England and Wales, having its registered office:
Edinburgh Gate, Harlow, Essex, CM20 2JE. Registered company number: 872828

www.heinemann.co.uk

Heinemann is the registered trademark of Pearson Education Limited

Text © University of York 2008

First published 1994
Second edition published 2000
This edition published 2008

12 11 10 09
10 9 8 7 6 5 4

British Library Cataloguing in Publication Data is available from the British Library on request.

ISBN 978 0 435631 47 5

Edited by Tony Clappison
Designed, produced, illustrated and typeset by Wearset Limited, Boldon, Tyne and Wear
Original illustrations © Pearson Education Limited 2008
Cover design by Wearset Limited, Boldon, Tyne and Wear
Picture research by Q2AMedia
Cover photo/illustration © NASA/Science Photo Library
Printed in China (SWTC/04)

Websites
The websites used in this book were correct and up-to-date at the time of publication. It is
essential for tutors to preview each website before using it in class so as to ensure that the URL
is still accurate, relevant and appropriate. We suggest that tutors bookmark useful websites and
consider enabling students to access them through the school/college intranet.

CONTENTS

ACKNOWLEDGEMENTS

The authors and publishers would like to thank the following individuals and organisations for permission to reproduce photographs:

p3 CBS Studios; **p4 TL** NASA/Jeff Hester and Paul Scowen Arizona State University; **p4 TR** Roger Bamba/Rex Features; **p4 BR** EFDA-JET; **p5 L** Royal Observatory Edinburgh/Science Photo Library; **p5 R** CXC, JPL-Caltech, J. Hester and A. Loll (Arizona State University)/NASA; **p9** The Trustees of the British Museum; **p10** Fortean Picture Library; **p12 L** Jim Sugar/Corbis; **p12 R** NASA/JPL; **p13** J.C. Revy/Science Photo Library; **p16** Ria Novosti/Science Photo Library; **p17** Hulton Archive/Getty Images; **p21** John Boden; **p23** Georgina Boyle; **p24** Markus Matzel/Das Fotoarchiv./Still Pictures; **p27** Simon Rendall/National Motor Museum; **p33** © BP p.l.c.; **p36 T** Peggy/Still Pictures; **p36 B** Peggy/Still Pictures; **p38** CEAM Foundation, Valencia (Spain); **p41** Document General Motors/Reuters/Corbis Sygma; **p42 T** Martin Bond/Science Photo Library; **p42 B** Julian Etchart/Still Pictures; **p43** Colin Cuthbert/Science Photo Library; **p44 T** Simon Fraser/Science Photo Library; **p44 MT** Corbis; **p44 MB** Corbis; **p44 B** Corbis; **p45** American Honda Motor Co; **p49 T** Bernhard Edmaier/Science Photo Library; **p49 B** B. Murton/Southampton Oceanography Centre/Science Photo Library; **p50** D.S. Kelley, University Of Washington; **p51** Lee Foster/Lonely Planet Images; **p54 T** James Holmes, Hays Chemicals/Science Photo Library; **p54 B** ELTECH Gruppo De Nora; **p56 L** Yoav Levy/Photolibrary; **p56 R** James King-Holmes/FTSS/Science Photo Library; **p57 L** Herve Berthoule/Jacana/Science Photo Library; **p57 R** Rich LaSalle/Getty Images; **p58** Barbara Strnadova/Science Photo Library; **p59 L** Sinclair Stammers/Science Photo Library; **p59 R** Zephyr/Science Photo Library; **p60 T** James Holmes, Hays Chemicals/Science Photo Library; **p60 BL** Patrick van Moer; **p60 BR** Stephen Sharnoff/Getty Images; **p62 L** Crown Copyright/Health & Safety Laboratory/Science Photo Library; **p62 R** Francoise Sauze/Science Photo Library; **p65** Ali Ender Birer/Shutter Stock; **p67 L** Pictorial Press Ltd/Alamy; **p67 R** Tom Hopkins/Getty Images; **p68** Karen Lucas; **p69** George Steinmetz/SPL/Photolibrary; **p72** Klaus Guldbrandsen/Science Photo Library; **p73** University of California; **p75 L** Simon Fraser/Science Photo Library; **p75 R** Faber Maunsell and Hugh Broughton Architects/British Antarctic Survey; **p76** NASA; **p77** Hobvias Sudoneighm; **p78** Philippe Plailly/Science Photo Library; **p82 T** Chris Gilbert/British Antarctic Survey; **p82 B** Pegaz/Alamy; **p83** Shutterstock; **p88** InterNetwork Media/Getty Images; **p89** Don Farrell/Getty Images; **p90** Fraser/Mauna Loa Observatory/SPL/Photolibrary; **p92** Andrew Syred/Science Photo Library; **p93** GeoStock/Getty Images; **p95 L** Altrendo Images/Getty Images; **p95 TR** CORBIS; **p95 BR** CORBIS; **p97** © National Archives and Records Administration; **p98** Greenshoots Communications/Alamy; **p99** Science Photo Library; **p100 L** Bettman/CORBIS; **p100 R** Getty Images; **p101 L** RMAX/Istockphoto; **p101 R** John Cole/Science Photo Library; **p102** Cordelia Molloy/Photolibrary; **p103 T** Gastromedia/Alamy; **p103 B** Dr Jeremy Burgess/Science Photo Library; **p104 TL** Eric von Michael/Alamy; **p104 BL** Adrian Thomas/Science Photo Library; **p104 TR** Buzz Pictures/Alamy; **p104 BR** Belkin

Thanks are due to the following for permission to reproduce copyright material:

p9 Stonehenge text: http://www.factsplusfacts.com; **p10** Dead Sea Scrolls text: http://www.factsplusfacts.com; **p10** Shroud of Turin text: http://www.factsplusfacts.com; **p17** figure 24: copyright P J Stewart, 2007 translated into electronic form by Carl Wenczek of Born Digital Ltd; **p35** table 5: Compendium of Experimental Cetane Number Data by M. Murphy, J. Taylor, and R. McCormick (NREL/SR-540-36805). This work was done by the National Renewable Energy Laboratory, a national laboratory of the U.S. Department of Energy; **p37** figure 22: GM and MJ Pilling 'What is photochemical smog?', Chemistry review 5:5, 1997, Philip Allan Publishers; **p37** figure 23: GM and MJ Pilling 'What is photochemical smog?', Chemistry Review 5:5, 1997, Philip Allan Publishers; **p50** associated text around figure 3: Reprinted with permission from Science News; **p60–62** ES6 Treasures of the Sea text: based on 'Amazing Organohalogens' by Gordon Gribble, American Scientist Vol. 92 pp. 342–347 on www.americanscientist.org; **p75** figure 13: Reprinted by permission from Macmillan Publishers Ltd: NATURE, May 1985; **p76** figure 16: Reprinted in part with permission from Environmental Science Technology Vol. 24, no. 4, p. 624, © (1991) American Chemical Society; **p78** figure 20: Chemistry of Atmospheres, 3rd edition, Richard Wayne (2000) Oxford University Press; **p79** figure 22: http://ozone.unep.org; **p80** figure 23: Copyright Guardian News & Media Ltd 1990; **p81** figure 24 a, b, c and d: R.G. Prinn (Massachusetts Institute of Technology), R.F. Weiss (Scripps Institution of Oceanography) and their colleagues in the NASA-supported Advanced Global Atmospheric Gases Experiment (AGAGE); **p86** figure 31: Kelter et al, 'Chemistry: A World of Choices', 1/e © 1999, McGraw-Hill. This material is reproduced with the permission of the McGraw-Hill companies; **p86** figure 32: NOAA Earth System Research Laboratory; **p87** figure 33: © Crown Copyright 2008, the Met Office; **p87** figure 34 a and b: Model simulations of average global temperature differences relative to the period 1901–1950, International Panel on Climate Change http://www.ipcc.ch; **p90** figure 38: Dave Keeling and Tim Whorf, Scripps Institute of Oceanography, NOAA Earth System Research Laboratory; **p92** figure 41: © Crown Copyright 2008, the Met Office

Every effort has been made to contact copyright holders of material reproduced in this book. Any omissions will be rectified in subsequent printings if notice is given to the publishers.

CONTRIBUTORS

The following people have contributed to the development of *Chemical Storylines AS* (Third Edition) for the Salters Advanced Chemistry Project:

Editors

Chris Otter (Project Director) — University of York Science Education Group (UYSEG)
Kay Stephenson — CLEAPSS

Associate Editors

Frank Harriss — Formerly Malvern College
Gwen Pilling — Formerly University of York Science Education Group (UYSEG)
Gill Saville — Dover Grammar School for Boys
David Waistnidge — King Edward VI College, Totnes
Ashley Wheway — Formerly Oakham School

Acknowledgement

We would like to thank the following for their advice and contribution to the development of these materials:

Sandra Wilmott (Project Administrator) — University of York Science Education Group (UYSEG)
Cheryl Alexander — University of York

Sponsors

THE SALTERS' INSTITUTE

We are grateful for sponsorship from the Salters' Institute, which has continued to support the Salters Advanced Chemistry Project and has enabled the development of these materials.

Dedication

This publication is dedicated to the memory of Don Ainley, a valued contributor to the development of the Salters Advanced Chemistry Project over the years.

The Third Edition Salters Advanced Chemistry course materials draw heavily upon the previous two editions and the work of all contributors, including the following:

First Edition

Central Team

George Burton	Cranleigh School and University of York
Margaret Ferguson (1990–1991)	King Edward VI School, Louth
John Holman (Project Director)	Watford Grammar School and University of York
Gwen Pilling	University of York
David Waddington	University of York

Associate Editors

Malcolm Churchill	Wycombe High School
Derek Denby	John Leggott Sixth Form College, Scunthorpe
Frank Harriss	Malvern College
Miranda Stephenson	Chemical Industry Education Centre
Brian Ratcliff	OCR (formerly Long Road Sixth Form College, Cambridge)
Ashley Wheway	Oakham School

Second Edition

Central Team

John Lazonby	University of York
Gwen Pilling (Project Director)	University of York
David Waddington	University of York

Associate Editors

Derek Denby	John Leggott College, Scunthorpe
John Dexter	The Trinity School, Nottingham
Margaret Ferguson	Lews Castle School, Stornoway
Frank Harriss	Malvern College
Gerald Keeling	Oundle School
Dave Newton	Greenhead College, Huddersfield
Brian Ratcliff	OCR (formerly Long Road Sixth Form College, Cambridge)
Mike Shipton	Oxted School, Surrey (formerly Reigate College)
Terri Vine	Loreto College (formerly Epsom and Ewell School)

INTRODUCTION FOR STUDENTS

The Salters Advanced Chemistry course for AS and A2 is made up of thirteen teaching modules. *Chemical Storylines AS* forms the backbone of the five AS teaching modules. There is a separate book of *Chemical Ideas*, and a *Support Pack* containing activities to accompany the AS teaching modules.

Each teaching module is driven by the storyline. You work through each storyline, making 'excursions' to activities and chemical ideas at appropriate points.

The storylines are broken down into numbered sections. You will find that there are **assignments** at intervals. These are designed to help you through each storyline and to check your understanding, and they are best done as you go along.

Excursions to Activities

As you work through each storyline, you will find that there are references to particular **activities**. Each activity is referred to at that point in the storyline to which it most closely relates. Activities are numbered to correspond with the relevant section of each storyline.

Excursions to Chemical Ideas

As you work through the storylines, you will also find that there are references to sections in *Chemical Ideas*. These sections cover the chemical principles that are needed to understand that particular part of the storyline, and you will probably need to study that section of the *Chemical Ideas* book before you can go much further.

As you study *Chemical Ideas* you will find **problems** relating to each section. These are designed to check and consolidate your understanding of the chemical principles involved.

Building up the Chemical Ideas

Salters Advanced Chemistry has been planned so that you build up your understanding of chemical ideas gradually. For example, the idea of chemical equilibrium is introduced in a simple, qualitative way in 'The Atmosphere' module. A more detailed, quantitative treatment is given in the A2 teaching modules 'Agriculture and Industry' and 'The Oceans'.

Sections in *Chemical Ideas* cover chemical principles that may be needed in more than one module of the course. As *Chemical Ideas* covers both AS and A2 content, those sections met for the first time at AS are clearly marked. The context of the chemistry for a particular module is dealt with in the storyline itself and in related activities. *Chemical Storylines* features coloured boxes carrying extra background chemistry (green boxes) and case studies or in-depth information about certain aspects of the storyline (purple boxes).

How much do you need to remember?

The specification for OCR Chemistry B (Salters) defines what you have to remember. Each teaching module includes at least one 'Check your knowledge and understanding' activity. These can be used to check that you have mastered all the required knowledge, understanding and skills for the module. Each 'Check your knowledge and understanding' activity lists whether a topic is covered in *Chemical Ideas*, *Chemical Storylines* or in the associated activities.

Salters Advanced Chemistry Project

ELEMENTS OF LIFE

Why a module on 'Elements of Life'?

This module tells the story of the elements of life – what they are, how they originated and how they can be detected and measured. It shows how studying the composition of stars can throw light on the formation of the elements that make up our own bodies and considers how these elements combine to form the 'molecules of life'. The module also takes the opportunity to look at some aspects of 'how science works'. In particular, developing models, identifying risk and benefit and seeing how the scientific community validates work.

The module begins with a journey through the Universe. Starting with deep space, the story unfolds through the galaxies, the stars and our own Sun and solar system. This section looks at the origin of the elements, introducing ideas about the structure of atoms, and briefly considers how elements combine to form compounds and the formation of molecules in the apparently inhospitable dense gas clouds of space, such molecules possibly being the origin of the 'molecules of life' which make up our bodies.

The second part of the module brings you back down to Earth! You learn how to measure amounts of elements (in terms of atoms) and so how to calculate chemical formulae. The story then leads into learning about patterns in the properties of elements and the Periodic Table.

Overview of chemical principles

In this module you will learn more about ideas you will probably have come across in your earlier studies:
- the Periodic Table
- protons, neutrons and electrons
- radioactivity and ionising radiation
- chemical bonding
- writing chemical equations
- the wave model of light
- the electromagnetic spectrum.
- relative atomic masses, relative molecular masses and relative formula masses
- chemical formulae
- ionic and covalent bonding

You will also learn new ideas about:
- amount of substance (moles and the Avogadro constant)
- nuclear fusion and nuclear equations
- the photon model of light
- atomic spectra
- the electronic structure of atoms
- shapes of molecules
- metallic bonding
- the relationship between structure and properties.

The chemical ideas about amount of substance, atomic structure, chemical bonding and the relationship between structure and properties are only *introduced* in this module. They will be consolidated and developed in later modules.

This technique of taking ideas only as far as you need to know to follow the module you are studying, and then building on them by repeating the process in later modules, is central to the Salters' approach to chemistry at this level.

ELEMENTS OF LIFE

EL1 *Where do the chemical elements come from?*

▲ **Figure 1** The Starship Enterprise – 'To boldly go …'

Space – the final frontier?

There are various theories about the origin of the Universe. The idea of the 'Big Bang' is still a front-runner for most cosmologists. Based on this theory, about three minutes after the 'Big Bang' the elements hydrogen and helium, along with small traces of lithium, began to form (or at least their nuclei did) from the huge, hot melting pot containing a myriad of tiny particles including protons and neutrons. The temperature was then about a hundred million degrees Celsius (10^8 °C)!

After about 10 000 years, the Universe had cooled sufficiently (to about ten thousand degrees Celsius (10^4 °C)) and electrons were moving slowly enough to be captured by oppositely charged protons. The Universe was (and still is) made up of mainly hydrogen and helium atoms.

The galactic empire

The Universe continued to cool and bits of dust and gas clumped together, pulled by their gravity, eventually forming gas clouds. The temperature of the gas clouds varied from 10 to 100 K (-263 to -173 °C). The particles had low kinetic energies and moved around relatively slowly, so that gravitational forces between the particles were able to keep them together.

Parts of the clouds gradually contracted in on themselves and the gases become compressed, forming 'clumps' of denser gas.

The densest part of the 'clump' is its centre. Here the gases are most compressed and become very hot, up to ten million degrees Celsius (10^7 °C). Such temperatures are high enough to trigger *nuclear* reactions. At these temperatures, atoms cannot retain their electrons and matter becomes a plasma of ionised atoms and unbound electrons.

A nuclear reaction is different to a chemical reaction. A chemical reaction involves the *rearrangement* of an atom's outer electrons, while a nuclear reaction involves a change in its nucleus. In a nuclear reaction *one element can change into another element* – something that would be impossible in a chemical reaction.

One nuclear reaction that takes place in the centre of 'clumps' is **fusion** – when lighter nuclei are fused together to form heavier nuclei. The nuclei need to approach each other at high speed, with a large kinetic energy, to overcome the repulsion between the positive charges on the two nuclei.

The nuclei of hydrogen atoms in the gas cloud join together by nuclear fusion, and the hydrogen turns into helium. The process releases vast quantities of energy, which causes the dense gas cloud to glow – the dense gas cloud has become a star (see Figure 5). Fusion is common in the centre of stars, where temperatures can reach hundreds of millions of degrees. Here are two examples of reactions which take place in the Sun:

$$^1_1H + {}^2_1H \rightarrow {}^3_2He + \gamma$$

$$^2_1H + {}^3_1H \rightarrow {}^4_2He + {}^1_0n$$

Notice that **atomic numbers** and **mass numbers** must balance in a nuclear equation.

After hydrogen, helium is still the most abundant element in space.

Other fusion reactions produce smaller amounts of heavier elements. Vast galaxies are formed where bright pinpoints of light show evidence of nuclear fusion starting – the first stars are beginning to shine.

These nuclear reactions also generate a hot wind which drives away some of the dust and gas, leaving behind the new stars. Planets that have condensed out of the remaining dust cloud often surround these stars.

▲ **Figure 2** 'The Pillars of Creation' – the 'fingers' emerging from the pillars of molecular hydrogen and dust contain small, very dense regions that are embryonic stars. This image was taken by the Hubble Space Telescope.

▲ **Figure 3** Starman? Are our origins in stardust? This image shows 70s pop icon David Bowie as his alter ego 'Ziggy Stardust'.

A star is born

Hydrogen is still the most common element in the Universe. Humans contain quite a lot of hydrogen, but we also contain other heavier elements as well. The theory of the evolution of the stars shows how heavy elements can be formed from lighter ones, and helps to explain the way elements are distributed throughout the Universe, including in ourselves.

The theory of how stars form is one of the major scientific achievements of the twentieth century. It was developed through observation of a range of stars at different stages in their development, studying them as they changed over time.

FUSION ON EARTH – A FUTURE 'GREEN' ENERGY SOURCE?

Forcing together the positive nuclei of atoms of light elements to create a heavier nucleus (and as a result a different element) requires extreme conditions of temperature and gravitational pressures, like those experienced in the Sun.

However, nuclear fusion releases large amounts of energy and no polluting emissions, so this could be a useful source of energy if such conditions were controllable on Earth.

JET

JET (Joint European Torus) is a tokamak, a machine in which strong magnetic fields are used to confine the plasma needed for such fusion reactions. In this class of device, the plasma chamber is doughnut shaped. The vessel is filled with gas at a very low pressure and this gas is converted to hot plasma by passing an electric current through it. The application of further strong magnetic fields keeps ('confines') the hot particles in the centre of the vessel and avoids melting of the walls. The JET tokamak is the largest in the world and is situated near Oxford.

JET: a bridge to ITER

ITER is a tokamak designed by an international team including Europe, Japan, China, India, the Russian Federation and the USA. The device's main aim will be to produce prolonged fusion power in a deuterium–tritium plasma. Scientific work at JET is now mainly devoted to testing out operating scenarios for ITER. It is hoped that ITER will be producing fusion power by around 2016.

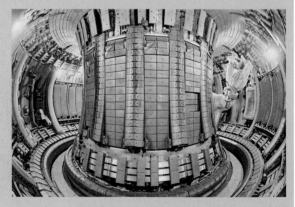

▲ **Figure 4** The JET machine seen from above.

Chemical Ideas 2.1 provides you with information about the structure of atoms and isotopes, while the first part of Chemical Ideas 2.2 explains nuclear reactions and their uses. These topics will help you with your study of this section.

Activity EL1.1 encourages you to explore the nature of science and how scientific models develop.

▲ **Figure 5** The glow of star formation in the Orion nebula.

Heavyweight stars

What happens next to a star depends on its mass. All stars turn hydrogen into helium by nuclear fusion. This process occurs fastest in the heaviest stars because their centres are the hottest and the most compressed. These **heavyweight** stars have very dramatic lives. The

temperatures and pressures at the centre of the star are so great that further fusion reactions take place to produce elements heavier than helium.

Layers of elements form within the star, with the heaviest elements near the centre where it is hottest and where the most advanced fusion can take place.

Figure 6 shows an example of the composition of the core of a typical heavyweight star after a few million years – long enough for extensive fusion to have taken place.

The element at the *centre of the core* is iron. When iron nuclei fuse they do not release energy but they *absorb* it. When the core of a heavyweight star reaches the stage where it contains mainly iron, it becomes unstable and explodes. These explosions are called **supernovae** – the most violent events in the Universe (Figure 7).

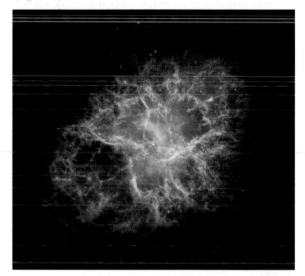

▲ **Figure 7** Violence in Crab nebula – scientists believe that such pictures are evidence of a supernova.

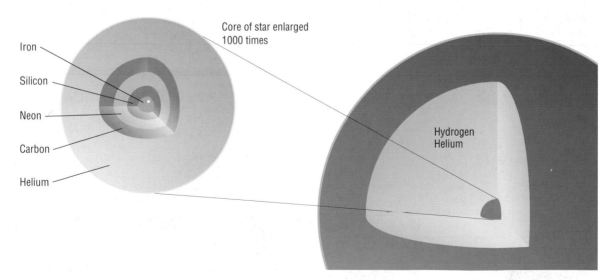

Core of star enlarged 1000 times

Iron
Silicon
Neon
Carbon
Helium

Hydrogen
Helium

▲ **Figure 6** A model of the core of a typical heavyweight star. Heavyweight stars vary in mass, but can be about 8 times the mass of our Sun.

As a result of a supernova, the elements in the star are dispersed into the Universe as clouds of dust and gas, and so the life cycle begins again.

Assignment 1

Identify the isotopes that are missing from the following nuclear equations.

a $^{12}_{6}C + ? \rightarrow ^{16}_{8}O$

b $^{14}_{7}N + ^{1}_{1}H \rightarrow ?$

c $^{7}_{3}Li + ? \rightarrow ^{4}_{2}He + ^{4}_{2}He$

Some types of meteorites contain large amounts of iron. **Activity EL1.2** gives you the opportunity to find out the concentration of iron in a sample of an iron compound.

The Sun – a lightweight among stars

The Sun is a **lightweight** star – it is not as hot as most other stars and will last longer than heavyweight stars. It will keep on shining until all the hydrogen has been used up and the core stops producing energy: there will be no supernova. Once the hydrogen is used up, the Sun will expand into a **red giant**, swallowing up the planets Mercury and Venus. The oceans on Earth will start to boil and eventually it too will be engulfed by the Sun. The good news for Earth is that the Sun still has an estimated 5000 million years' supply of hydrogen left!

As red giants get bigger they also become unstable and the outer gases drift off into space, leaving behind a small core called a **white dwarf**, about one-hundredth of the size of the original star.

How do we know so much about outer space?

The work of chemists has made a vital contribution to the understanding of the origin, structure and composition of our Universe. To do this, they have used a method called **spectroscopy**.

Many different spectroscopic techniques exist (others are discussed in **Polymer Revolution**, **Colour by Design** and **Medicines by Design** modules), but all are based on one very important scientific principle – under the right conditions a substance can be made to **absorb** (take in) or **emit** (give out) electromagnetic radiation in a way that is characteristic of that substance. The **electromagnetic spectrum** in Figure 8 shows the different types of electromagnetic radiation.

If we analyse this electromagnetic radiation (such as ultraviolet light, visible light or radio waves) we can learn a lot about a substance. Sometimes we just want to know what it is. At other times we want to find out very detailed information about it, such as its structure and the way its atoms are held together. Figure 9 shows how visible light can be analysed using a spectrograph.

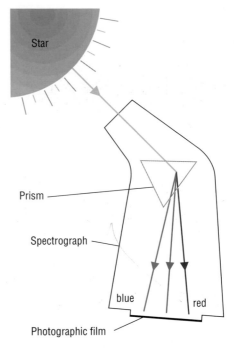

▲ **Figure 9** The frequencies in a beam of light can be analysed using a spectrograph.

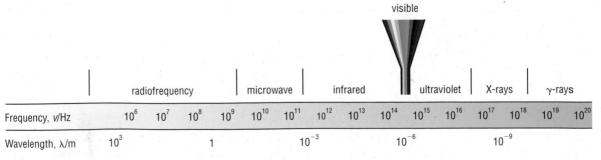

▲ **Figure 8** The electromagnetic spectrum.

Absorption spectra

The glowing regions of all stars emit light of all frequencies between the ultraviolet and the infrared parts of the electromagnetic spectrum. The Sun emits mainly visible light – its surface (**photosphere**) glows like an object at about 6000 K. Some stars are cooler than the Sun; others are much hotter, reaching temperatures as high as 40 000 K and emitting mainly ultraviolet radiation.

Outside the star's photosphere is a region called the **chromosphere** (Figure 10). The chromosphere contains ions, atoms and, in cooler stars, small molecules. These particles absorb some of the light that is emitted from the glowing photosphere. So when we analyse the light which reaches us from the star, we see that certain frequencies are missing – the ones which have been *absorbed*.

Further out still is the **corona**. Here the temperature is so high that the atoms have lost many of their electrons. For example, Mg^{11+} and Fe^{15+} ions have been detected.

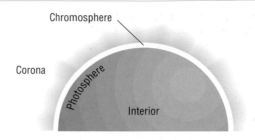

▲ **Figure 10** The structure of the Sun, a typical star.

β Centauri is a B-type star (a type of very hot star). The spectrum of the visible light reaching us from β Centauri, the star's visible **absorption spectrum**, is illustrated in Figure 11. You can clearly see the **absorption lines** – they appear as black lines on the bright background of light emitted from the star because these correspond to the frequencies that are missing.

The absorption lines in β Centauri's spectrum arise only from hydrogen atoms and helium atoms. These

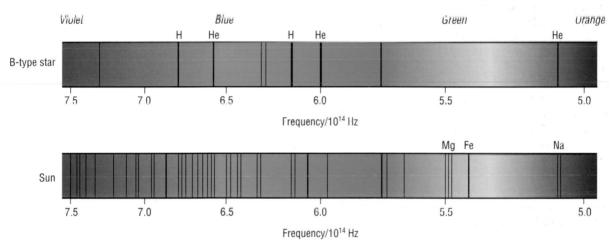

▲ **Figure 11** Absorption spectra of a B-type star (e.g. β Centauri) and the Sun. Black lines occur where frequencies are missing from the otherwise continuous spectra. (Note that here the frequency increases from right to left, which is a common convention for absorption spectra.)

Assignment 2

The spectrum of the light received on Earth from Sirius A is shown in Figure 12. Sirius is an A-type star, which is less hot than a B-type star.

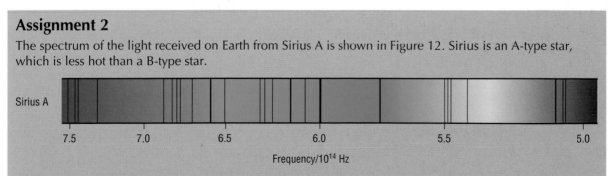

Figure 12 The absorption spectrum of the star Sirius A.

Compare this spectrum with those in Figure 11 and name *five* elements which the spectrum shows are present in Sirius A.

are the only atoms able to absorb visible light at the very high temperatures of β Centauri.

For comparison, the Sun's absorption spectrum is also shown in Figure 11. Because the Sun is at a lower temperature, different particles are able to absorb visible light. For example, lines from sodium, iron and magnesium can be seen. The Sun's chromosphere consists mainly of hydrogen and helium but, at the temperature of the Sun, these do not absorb visible light.

Emission spectra

When the atoms, molecules and ions around stars absorb electromagnetic radiation, they are raised to higher energy states called **excited states**. The particles can lose their extra energy by emitting radiation. The resulting **emission spectra** can also be detected on Earth.

During a total solar eclipse, the glow of the Sun's photosphere is completely blocked out by the Moon. The light being *emitted* by the chromosphere is all that can be seen, and it is then that the presence of hydrogen and helium is revealed. Hydrogen atoms dominate the chromosphere's *emission* spectrum, but a helium emission line can also be seen. The hydrogen emission spectrum is shown in Figure 13. Helium gets its name from *helios*, the Greek word for the Sun. This previously unknown element was first detected in the chromosphere during the eclipse of 1868.

Careful and detailed study of all the types of radiation received on Earth from outer space has allowed a picture to be built up of part of the chemical composition of the Universe. More recently, it has been possible to add to this picture by sending space probes (such as those used in the Voyager missions) fitted with a variety of spectroscopic devices.

Chemical Ideas 6.1 will help you to find out more about the information chemists can obtain from spectra. This information includes the arrangements of electrons in shells, which is described in **Chemical Ideas 2.3**.

In **Activity EL1.3** you can look at the light emitted when compounds of certain elements are heated in a Bunsen flame.

Our solar system

Our solar system probably condensed from a huge gas cloud that gradually contracted under the force of gravity. As rings of gas and dust condensed around the Sun, the planets were formed. This material originated from a supernova and therefore contained a range of elements. The non-volatile elements condensed near to the Sun, where temperatures were greatest, while the more volatile elements condensed further away from the Sun at lower temperatures.

Our solar system is therefore made up of small, dense, rocky planets close to the Sun, and giant fluid planets further away from the Sun. Conditions on all the planets are very different and some of their chemistry seems very unusual when compared with our experiences on Earth. Figure 14 illustrates how unusual some of the chemistry is.

Does this mean that the composition of the Earth is fixed to just that blend of elements that condensed around the Sun all those billions of years ago? The answer is no. Some of the atoms which formed the Earth were unstable, and began breaking down into atoms of other elements by **radioactive decay**. This process is still going on today.

Now we are also able to produce our own unstable atoms and have learned how to use radioactive decay processes to our benefit. Radioactive tracers can be used in the body. You will find out about this later in the module (**Section EL3**). Radioactive isotopes can also be used to date archaeological and geological findings.

You can remind yourself about radioactive decay, as well as learning about how half-lives can be used for dating purposes, in the last section of **Chemical Ideas 2.2**.

Activity EL1.4 uses pasta to illustrate the ideas of radioactive decay and half-life!

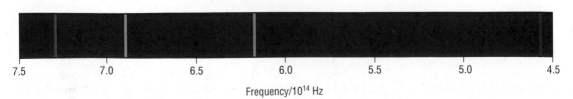

7.5   7.0   6.5   6.0   5.5   5.0   4.5

Frequency/10^{14} Hz

▲ **Figure 13** The hydrogen emission spectrum in the visible region.

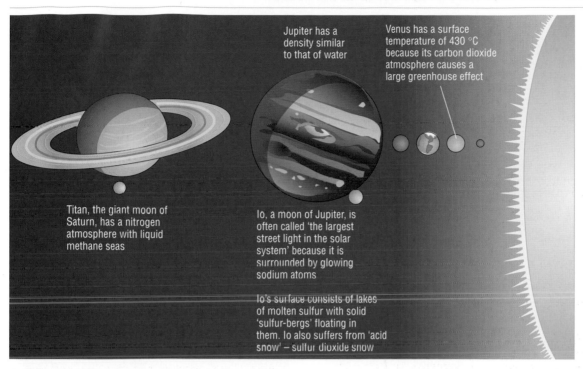

Jupiter has a density similar to that of water

Venus has a surface temperature of 430 °C because its carbon dioxide atmosphere causes a large greenhouse effect

Titan, the giant moon of Saturn, has a nitrogen atmosphere with liquid methane seas

Io, a moon of Jupiter, is often called 'the largest street light in the solar system' because it is surrounded by glowing sodium atoms

Io's surface consists of lakes of molten sulfur with solid 'sulfur-bergs' floating in them. Io also suffers from 'acid snow' – sulfur dioxide snow

▲ **Figure 14** Did you know...

ARCHAEOLOGICAL USES OF CARBON-14 DATING – SUCCESSES AND A NOTE OF CAUTION!

The relatively short half-life (compared to other radioisotopes) of just less than 6000 years means that C 14 has been used extensively to back up other dating methods used for archaeological artefacts. Some notable examples are described below.

Stonehenge

Archaeologists have come to realise that Stonehenge was built in three stages. The first stage was a circle of wooden timbers surrounded by a ditch. Excavations of the ditch revealed many animal bones and deer antlers that had been buried there. Carbon-14 dating of this material has revealed that the first circle of Stonehenge was constructed in about 3100 BC.

Lindow Man

The body of the man shown in Figure 15 was discovered in August 1984 when workmen were cutting peat at Lindow Moss bog in north west England. Research at the British Museum has allowed us to learn more about this person – his health, his appearance and how he might have died – than any other prehistoric person discovered in Britain.

The conditions in the peat bog meant that the man's skin, hair and many of his internal organs are well preserved. Radiocarbon dating shows that he died between AD 20 and 90.

▲ **Figure 15** Lindow Man suffered a nasty death!

The Iceman

The Iceman is a frozen body found in northern Italy in 1991. Samples of his bones, boot, leather and hair were dated. The results showed that he lived almost five and a half thousand years ago (3300–3100 BC), during the age when people first began using copper in Europe. Radiocarbon dating was tremendously important in dating the Iceman.

Dead Sea Scrolls

The Dead Sea Scrolls are thought to be remnants of a library that had been wrapped in linen cloth and stored in jars in caves among the cliffs of Qumran near Jerusalem. The first scrolls were discovered in 1947 when an Arab herdsman came across them while searching for a stray goat. About 800 scrolls in Hebrew and Aramaic have been found, the latest in 1956. They include the oldest known manuscripts from the Old Testament. Carbon-14 dating has been used to confirm the dates of these scrolls to be between 150 and 5 BC.

The Shroud of Turin

The Shroud of Turin (Figure 16) is a religious artefact believed by many to be the burial shroud of Jesus. However, in 1988 carbon-14 dating was undertaken on a sample of the shroud.

Nature, the international weekly journal of science, published an article about the carbon-14 dating co-authored by 21 scientists from the University of Oxford, the University of Arizona, the Institut für Mittelenergiephysik in Zurich, Columbia University and the British Museum. The conclusion according to the *Nature* article was clear – the analysed linen was thought to date from between AD 1260 and 1390.

However, new information published in several scientific papers in early 2004 seemed to show that the Shroud of Turin had not been successfully carbon-14 dated. It seems that what was tested was nothing more than a mixture of old thread and new thread from a medieval patch (historical documentation suggests that the repair may have been made in 1530 or 1531).

There had been nothing wrong with the rigour of the initial testing, but it just goes to show that you are only as good as the material you are given!

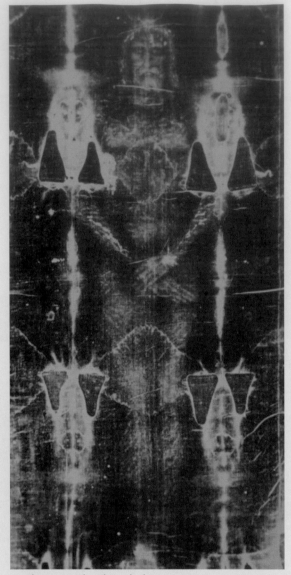

▲ **Figure 16** The Shroud of Turin.

EL2 *The molecules of life*

The dense (or molecular) gas clouds

Although hydrogen is the most common element in space, its atoms are relatively few and far between. There is about one atom per cubic centimetre (cm^3) in the space between the stars, compared with more than 1×10^{19} atoms per cm^3 in the air you are breathing now. With a density of hydrogen atoms in space as low as this, there is almost no chance that hydrogen atoms will come together to form hydrogen molecules.

However, there are some regions between the stars where molecules do form. These are called **dense gas clouds** or **molecular gas clouds**, though they are hardly dense by standards on Earth. These regions may contain as few as 100 particles per cm^3 up to as many as 1×10^6 particles per cm^3. This sort of density means that there may be distances between the particles many millions of times the size of the particles themselves. The gas clouds are made up of a mixture of atoms and molecules, mainly of hydrogen, together with a **dust** of solid material from the break up of old stars.

Cold chemistry and the 'molecules of life'

You have heard a lot about atoms so far, but humans are made up of molecules and some ions, rather than single atoms. So what are the molecules of life and how did they come into existence?

Molecules are formed in the colder parts of the Universe when individual atoms happen to meet and bond to one another. (Molecules do not exist in stars because the bonds connecting the atoms cannot survive at the high temperatures there.) Molecules and fragments of molecules have been detected in dense gas clouds, both by radio and infrared telescopes on Earth and by spectroscopic instruments carried by rockets.

Table 1 shows some of the chemical species found in dense gas clouds – some will look familiar, but many will look strange.

Many of the substances in Table 1 can be described as **organic** species — this means that they contain carbon atoms bonded to elements other than just oxygen.

There is something familiar about the elements in these species – they are the elements which are the major constituents of the human body.

Chemical Ideas 3.1 tells you about the ways in which elements can combine with each other. Try testing your understanding in **Activity EL2.1** Why do atoms form ions?

Molecules (and ions) that contain covalent bonds take up definite shapes. You can read about the shapes of molecules in **Chemical Ideas 3.2**.

Balloons can be used to give good illustrations of molecular shapes. You can try using them like this and then use card sorting to focus your knowledge of molecular shapes in **Activity EL2.2**.

Where did the molecules of life come from?

Some scientists have suggested that the molecules in the dense gas clouds were the building blocks which reacted together to make the molecules that form the basis of life on Earth. They believe that the energy needed to make these reactions take place came from ultraviolet radiation, X-rays and cosmic rays, and closer to Earth from lightning flashes.

In 1950 an American scientist, Stanley Miller, put methane (CH_4), ammonia (NH_3), carbon dioxide (CO_2) and water – simple molecules like those present in the dense gas clouds – into a flask and heated them (Figure 17, page 12). He also subjected the mixture to an electrical discharge to simulate the effect of lightning. On analysing the products, Miller found that some of the reaction mixture had been converted into amino acids. Amino acids are the building blocks of proteins – proteins are a group of compounds needed for correct cell functioning.

In a separate experiment another scientist in the US, Leslie Orgel, made a very dilute solution of ammonia and hydrogen cyanide (HCN) and froze it for several days. When he analysed the 'ice', he identified amino acids and the compound adenine. Adenine is one of four compounds which, together with phosphate units and a sugar called deoxyribose, make up DNA (deoxyribonucleic acid) – the substance which contains the genetic code for reproduction.

Both Miller and Orgel showed that molecules like those in the dense gas clouds could react together

Table 1 Some chemical species in the dense gas clouds.

Monatomic	Diatomic	Triatomic	Tetra-atomic	Penta-atomic
C^+	H_2	H_2O	NH_3	HCOOH
Ca^{2+}	OH	H_2S	H_2CO	NH_2CN
H^+	CO	HCN	HNCO	HC_3N
	CN	HNC	HNCS	C_4H
	CS	SO_2	C_3N	CH_2NH
	NS	OCS		CH_4
	SO	N_2H^+		
	SiO	HCS^+		
	SiS	HCO^+		
	C_2	NaOH		
	CH^+			
	NO			

Hexa-atomic	Hepta-atomic	Octa-atomic	Nona-atomic	Others
CH_3OH	CH_3CHO	$HCOOCH_3$	CH_3CH_2OH	HC_9N
NH_2CHO	CH_3NH_2		CH_3OCH_3	$HC_{11}N$
CH_3CN	H_2CCHCN		CH_3CH_2CN	
CH_3SH	CH_3C_2H		HC_7N	
CH_2CCH				

▲ **Figure 17** Stanley Miller used this apparatus to make amino acids from simple molecules.

STARDUST – AN UPDATE

Many scientists believe that comets are largely made of the original material from which our solar system formed and may contain pre-solar system interstellar grains. According to scientists, continued analysis of these celestial specks may well yield important insights into the evolution of the Sun and its planets, and possibly even the origin of life.

In January 2004, the Stardust spacecraft actually flew through comet dust from Wild 2 and captured specks of it in a very light, low-density substance called aerogel. Stardust's return capsule parachuted to the Utah Test and Training Range on 15th January, 2006 – seven years after its launch.

Analysis of the dust has indicated that it contains polycyclic aromatic hydrocarbons (PAHs) which are molecules made of carbon and hydrogen that are common in interstellar space – and in barbecue grill soot. Studies also show the presence of organic compounds very rich in oxygen and nitrogen.

These molecules are of particular interest to astrobiologists because these kinds of compounds play important roles in terrestrial biochemistry. Could these relatively simple molecules go on to form larger molecules and ultimately biological molecules, such as amino acids and the bases that link together to form DNA? The evidence is building but the jury is still out!

under conditions similar to those that existed during the early history of the Earth to form some of the molecules of life. These experiments gave added weight to the suggestion that life on Earth has its origin in molecules from outer space.

There are other theories on how molecules of life may be formed in outer space, theories that may be supported by the space mission *Stardust*, launched by NASA in 1999. One of its aims is to collect a sample from a dense gas cloud, but in the meantime *Stardust* has already intercepted the comet Wild 2 and sampled some of the material in the comet's tail (see Figure 18 in the box, right).

One theory suggests that the key is held by the interstellar dust thrown out by stars, which is protected from the most intense ultraviolet radiation in space by the dense gas clouds. This dust contains minute dust particles. At the centre of each particle is a hard core made up of giant structures such as graphite (carbon), silica (silicon dioxide), iron and other substances. Around this core are solid compounds with simple structures, such as water, ammonia, methane and carbon dioxide. The properties of such compounds vary greatly depending on their structure.

▲ **Figure 18** An artist's impression of Stardust encountering Wild 2.

You can remind yourself of simple ideas of structure and bonding by revisiting **Chemical Ideas 3.1**. This concept will be revisited frequently as you continue studying this course.

Activity EL2.3 will help you to relate the properties of a substance to the arrangement of its constituent particles.

The temperatures in gas clouds are too low for 'normal' chemical reactions to occur. However, the ultraviolet light which does penetrate the clouds can break the covalent bonds in the simple molecules and so reactions can take place, leading to the formation of larger molecules. These molecules, in turn, can react at slightly higher temperatures to form biological compounds such as amino acids and the bases that link together to form DNA.

These theories are being tested in a series of experiments using the *Stardust* spacecraft. As Scott Sandford, one of its leaders, pointed out back in 1999: "Even if we only find three grains, they will be the only three we have – all that science has to study." *Stardust*'s rendezvous with Wild 2 has given science a good start!

You will take a more detailed look at proteins, DNA and other 'molecules of life' later in the course in the **Engineering Proteins** module.

Activity EL2.4 will help you check the notes you have made on **Sections EL1** and **EL2**.

EL3 *What are we made of?*
Elements and the body

If you asked a number of people the question: 'What are you made of?' you would get a variety of different answers. Some people would use biological terms and talk about organs, bones and so on. Others might answer in more detail and mention proteins, fats and DNA. A chemist would be most likely to talk about atoms and molecules, or elements and compounds.

As a chemist, you know that you are not really made up of a mixture of elements but rather a mixture of compounds, many of which appear quite complicated. You will be finding out more about some of these compounds later in the module. To begin with, however, you will look at the elements that are most likely to be in the compounds in your body.

Elements in the body are classified as one of three types:

▲ **Figure 19** A model of the haemoglobin molecule – a biological molecule essential to life.

- **major constituent elements**, which make up 2–60% of all the atoms present; these are hydrogen, oxygen, carbon and nitrogen
- **trace elements**, which make up 0.01–1%, e.g. calcium and phosphorus
- **ultra-trace elements**, which make up less than 0.01%, e.g. iron and iodine.

Our health depends on the presence of these elements in the correct amounts. People with osteoporosis tend to suffer from frequent bone fractures and curvature of the spine. Osteoporosis is a condition resulting from a lack of calcium incorporated into the bone structure. The condition is usually observed in women who have passed through the menopause.

An increased intake of calcium can help the condition – scientists wanted to find out how much of the calcium which is eaten is taken up by the bloodstream and becomes part of the bones. One way is to use a tracer technique. Patients are given a meal containing a radioactive calcium isotope, and the amount of radioactive calcium which is absorbed into the bloodstream from the gut is measured.

Assignment 3

a What sort of measurements do you think it would be necessary to make on patients if they received treatment involving a radioactive tracer?

b Suggest why there are advantages in using a short-lived radioactive isotope of calcium in these types of treatment.

c Suggest what might be the disadvantages of using a short-lived isotope in these types of treatment.

You have already read about the structure of the atom, the occurrence of isotopes and radioactive decay in **Chemical Ideas 2.1** and **2.2**.

Radioactive isotopes can be dangerous and an alternative method is to analyse for strontium, an element very similar to calcium (they are both in Group 2 of the Periodic Table). During bone formation strontium-90 (a non-radioactive isotope) is taken up into bones in a similar way to calcium.

Normal, non-radioactive strontium can be analysed in the presence of calcium because strontium compounds emit a characteristic red light when they are placed in a flame. (You can learn about emission spectroscopy in **Chemical Ideas 6.1**.) The intensity of this red light is a measure of how much strontium is present. The amount of strontium present in a bone sample is proportional to the amount of calcium present. The ability to monitor patterns of absorption of calcium through the use of non-radioactive strontium has led to a greater understanding of calcium deficiencies and improvements in the treatment of bone disorders such as osteoporosis.

The calcium content of bones can now be monitored routinely in hospitals. The procedure is quick and accurate and involves measuring bone density by comparing the transmission of X-rays through bone and soft tissue.

Counting atoms of elements

Table 2 lists the masses and proportions of the major constituent elements in a person of average mass (about 60 kg).

Table 2 The major constituent elements in the human body.

Element	Mass in a 60 kg person/g	Percentage of atoms
oxygen	38 800	25.5
carbon	10 900	9.5
hydrogen	5 990	63.0
nitrogen	1 860	1.4

Notice that Table 2 gives conflicting evidence about the importance of different elements in the body. For example, there are more atoms of hydrogen in your body than atoms of any other element, but hydrogen contributes far less than carbon or oxygen to the mass of your body. So a list of masses alone does not allow us to decide which of the three categories of 'elements of life' an element belongs to. To do this, we need to determine *how many atoms* of each element there are.

Chemists can convert masses of elements into a measure of the number of atoms they contain by making use of **moles**. When we are dealing with elements, 1 mole is the amount of an element which contains the same number of atoms as 12 g of carbon.

Because atoms have exceedingly small masses, the number of atoms in 1 mole of an element is large. In fact it is very, very large – approximately 6×10^{23} atoms per mole. There are almost 1000 moles of carbon in a 60 kg person, so there are an awful lot of carbon atoms in the human body.

Once the number of moles of atoms of each element in the body has been calculated from its mass, these can be added to give the total number of atoms in the body. The percentage of atoms of each element in the body can then be worked out using the total.

You can make and analyse a Group 2 compound in **Activity EL3**.

Chemical Ideas 1.1 tells you more about moles and how to use them in calculations.

Table 2 shows that the ratio of hydrogen atoms to oxygen atoms in your body is almost 2 : 1. That's because 65% of the mass of your body is water and the chemical formula of water is H_2O. You have probably known that formula for a long time. But how did you know it? Did you ever work it out? It's not difficult – provided that you know about moles.

Assignment 4

Putting information in a table like Table 2 is often not the most striking way to present it. Pie charts or bar charts, for example, can be better.

Draw a pie chart to represent the proportions by mass of the four major constituent elements in the body. Label the fifth 'slice' of the chart to represent the contribution of all the trace and ultra-trace elements.

Now draw another pie chart, this time to represent the percentages of atoms.

How do you think that pie charts provide a better way of seeing and comparing the information in this case?

A trace is all you need

You saw earlier in this section that about 99% of your body is made up of atoms of only four elements – hydrogen, oxygen, carbon and nitrogen. These elements are obviously vital for life. But the trace and ultra-trace elements, although they make up only the remaining 1% or so of the body, are also essential for good health.

Table 3 lists the proportions and functions of the trace elements in a person. The ultra-trace elements – which include cobalt, copper, iodine, iron, manganese, molybdenum, silicon, vanadium and zinc – are not given because their quantities are so small.

Table 3 Trace elements in the human body.

Element	Mass in a 60 kg person/g	Percentage of atoms	Function
calcium	1200	0.31	major component of bone; required in some enzymes
phosphorus	650	0.22	essential for the synthesis of chemicals in the body and for energy transfer
potassium	220	0.06	required for correct functioning of the nervous system
sulfur	150	0.05	required in proteins and other compounds
chlorine	100	0.03	needed for regulation of osmotic balance
sodium	70	0.03	required for correct functioning of the nervous system
magnesium	20	0.01	needed for regulation of enzyme reactions and crucial for good bone structure

EL4 *Looking for patterns in elements*

When the elements were being discovered, and more being learned about their properties, chemists looked for patterns in the information they had assembled.

For example, there are close similarities between calcium and strontium. Magnesium and barium can be added to them to make a 'family' or 'group' of four elements. Your earlier studies probably introduced you to two other 'families' – lithium, sodium and potassium, and then fluorine, chlorine, bromine and iodine.

Activity EL4.1 shows the similarities in the properties of the Group 2 elements. You will need to write simple balanced equations in this activity. If you feel in need of revision, **Chemical Ideas 1.2** will help.

Chemical Ideas 11.2 looks at Groups 1 and 2 in the Periodic Table and will allow you to check your results from **Activity EL4.1**.

Fifty-nine of the ninety-two naturally occurring elements had been discovered by 1850, so the search for patterns among the elements was particularly fruitful in the mid-nineteenth century.

Much of the work was done by Johann Döbereiner and Lothar Meyer in Germany, John Newlands in England and Dmitri Mendeleev in Russia. These chemists looked at similarities in the chemical reactions of the elements they knew about, and also patterns in physical properties such as melting point, boiling point and density.

Mendeleev's groupings (see Figure 20) were seen as the most credible. Elements were arranged in order of increasing atomic masses, so that elements with similar properties came in the same vertical group. However, Mendeleev reversed the positions of iodine and tellurium because it made more sense for iodine to be grouped with chlorine and bromine.

Mendeleev's values for atomic masses were not accurate because the existence of isotopes was not known at that time. In **Activity EL4.2** you can use data from a mass spectrometer to find the mass numbers of isotopes and work out the relative atomic mass of an element.

	Group I	Group II	Group III	Group IV	Group V	Group VI	Group VII	Group VIII
Period 1	H							
Period 2	Li	Be	B	C	N	O	F	
Period 3	Na	Mg	Al	Si	P	S	Cl	
Period 4	K Cu	Ca Zn	* *	Ti *	V As	Cr Se	Mn Br	Fe, Co, Ni
Period 5	Rb Ag	Sr Cd	Y In	Zr Sn	Nb Sb	Mo Te	* I	Ru, Rh, Pd

▲ **Figure 20** A form of Mendeleev's Periodic Table – the asterisks denote elements which he thought were yet to be discovered.

Chemical Ideas 6.5 explains mass spectrometry and how a 'time of flight' mass spectrometer works.

Also, unlike Newlands, Mendeleev left gaps in his table of elements. These gaps were very important because they allowed for the discovery of new elements. This shows an awareness of the changing nature of scientific knowledge.

Mendeleev was so confident of the basis on which he had drawn up his table that he made predictions about elements yet to be discovered. In 1871, he predicted the properties of an element he called **eka-silicon**, which he was confident would eventually be discovered to fill the gap between silicon and tin in his Periodic Table. Mendeleev's predictions are shown in Table 4. The missing element was discovered in 1886 and called **germanium**. Its properties are in excellent agreement with Mendeleev's predictions. Later work had validated Mendeleev's earlier ideas.

Table 4 Mendeleev's predictions for the properties of eka-silicon. Researching the properties of germanium will allow you to see how close Mendeleev was in his predictions.

Property	Prediction
appearance	dark-grey solid
relative atomic mass	72
density	$5.5\,\mathrm{g\,cm^{-3}}$
reaction with water	none
reaction with acid	very little
reaction with alkali	more than with acid
oxide	basic, reacts with acid
chloride	liquid, boiling point <100 °C

Since Mendeleev's death in 1907, eight elements have been discovered and many more have been made in the laboratory. The first two elements to be made synthetically were neptunium ($Z = 93$) and plutonium ($Z = 94$). They were formed by bombarding uranium with neutrons. By 1961, elements up to $Z = 103$ had been made. By 2007, the heaviest element synthesised had an atomic mass of 118. Unfortunately it only existed for 200 microseconds!

The modern Periodic Table is based on the one originally drawn up by Mendeleev. It is one of the most amazingly compact stores of information ever produced – with a copy of the Periodic Table in front of you, and some knowledge of how it was put together, you have thousands of facts at your fingertips!

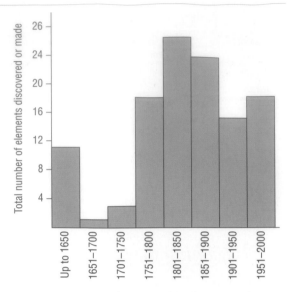

▲ **Figure 21** Historical pattern of the discovery of elements – many were discovered in the nineteenth century. Many 'artificial' elements have been made since the 1950s. You might like to search the Internet to see if you can explain jumps in discovery, such as that occurring after 1750.

▲ **Figure 22** Dmitri Mendeleev's (1834–1907) ideas form the basis of our modern classification of the elements.

Chemical Ideas 11.1 tells you more about the modern Periodic Table.

Activity EL4.3 allows you to use the Internet to look for patterns in the properties of elements.

THE SEARCH FOR 'ARTIFICIAL' ELEMENTS

Glenn Seaborg won the Nobel Prize for Chemistry in 1951 for his discoveries in the chemistry of the 'artificial' elements beyond uranium. He was co-discoverer of plutonium and all the elements from plutonium to nobelium (Z = 102). He thought his greatest honour was the naming of element 106 after him (seaborgium, Sg). He was the first living scientist to be so remembered.

▷ **Figure 23** Seaborgium was named after this man, Glenn Seaborg.

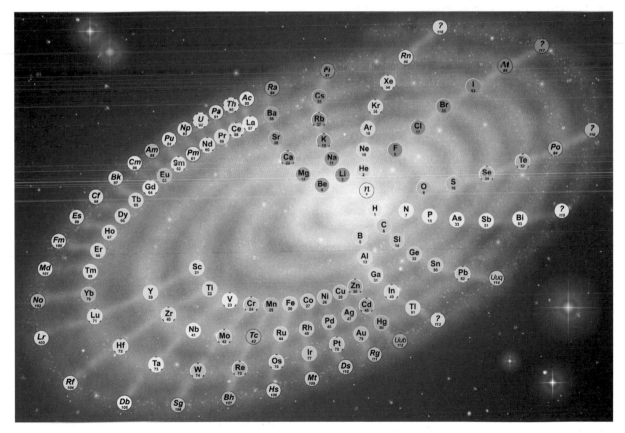

▲ **Figure 24** Connecting the two main sections of the **Elements of Life**? A galactic Periodic Table!

EL5 *Summary*

You began this module by considering the origin of the elements in stars. This allowed you to revise and develop your ideas about atomic structure and the use of scientific models. You learned to write nuclear equations for the reactions taking place, as well as how radioactive decay processes can be used to benefit society. You learned about atomic spectra and how spectroscopy is used to tell us about the composition of the Universe.

You then went on to study the ways that chemical elements combine together to form compounds, and saw how the chemical elements in outer space could have combined to form the 'molecules of life'.

The module finished by looking at the elements present in human beings. To work out the proportions of the different kinds of atoms in a human body, you need to know about moles of atoms. You learned how to convert the masses of elements that combine into moles of atoms, and so work out chemical formulae. You also learned how to use chemical formulae to write balanced chemical equations.

Having come across a number of different elements, you then found out about the ways in which chemists developed a system for classifying those elements and you looked in detail at the modern Periodic Table. You also went on to look at the chemistry of Group 2 elements.

Activity EL5 will help you check the notes you have made on **Sections EL3** to **EL5**.

DEVELOPING FUELS

Why a module on 'Developing Fuels'?

This module tells the story of petrol and diesel: what they are and how they are made. It also describes the work of chemists on improving fuels for motor vehicles, and in searching for and developing alternative fuels and sustainable energy sources for the future. Important ideas about vehicle pollutants and their control are also covered.

Some fundamental chemistry is introduced to achieve this. There are two main areas. First, it is important to understand where the energy comes from when a fuel burns. This leads to a study of enthalpy changes in chemical reactions, the use of energy cycles and the relationship between energy changes and the making and breaking of chemical bonds. Second, the module provides an introduction to organic chemistry. Alkanes are studied in detail and other homologous series, such as alcohols and ethers, are mentioned.

Isomerism is looked at in connection with the octane numbers of petrol, and simple ideas about catalysis arise out of the use of catalytic converters to control exhaust emissions. There is also a brief qualitative introduction to entropy which follows from a consideration of why the liquid components of a petrol-blend mix together. All these topics will be developed and used in later modules.

Overview of chemical principles

In this module you will learn more about ideas you will probably have come across in your earlier studies:
■ balancing equations
■ simple organic chemistry and homologous series
■ useful products from crude oil
■ combustion of alkanes
■ exothermic reactions
■ catalysis.

You will learn more about some ideas introduced in the Elements of Life module:
■ moles
■ empirical and molecular formulae
■ covalent bonding
■ polar bonds
■ molecular shape.

You will also learn new ideas about:
■ calculating reacting quantities using balanced chemical equations
■ enthalpy changes and enthalpy cycles
■ bond enthalpies
■ nomenclature of organic compounds
■ properties of alkanes
■ isomerism
■ alcohols and ethers
■ entropy
■ molar volumes of gases.

DF

DEVELOPING FUELS

DF1 *The vehicle of the future?*

Why don't we all drive electric vehicles? After all, they are clean, quiet and cause no exhaust pollution. Until recently electric vehicles were associated with poor performance. This was because of their heavy batteries which took a long time to recharge. Typically an electric vehicle had to be recharged overnight for several hours. Electrical energy trickled into the batteries at a rate of about 55 joules per second (55 W).

▲ **Figure 1** An electric car for the future – the Tesla roadster, photographed in 2007.

Compare this with the rate at which you can fill up a petrol-engined car with its source of energy. A petrol pump delivers petrol at about 1 litre per second, and a litre of petrol transfers 34 000 000 J of energy when you burn it. So a petrol pump replaces the car's energy supply over 600 000 times faster than a battery charger. These numbers are illustrated in a different way in Figure 2.

Petrol is a highly concentrated energy source. Couple this with the ease and cheapness of building petrol engines, and the fact that petrol stations and car repair centres are already established, and it's not surprising that petrol vehicles are so popular. However, current advances in technology have led to high-performance electric vehicles. For example, the Tesla roadster has a top speed of 130 mph and recharges fully in three hours (2007 data). Add to this the fact that electric cars are exempt from the road fund license and congestion charging and you can see that the number of electric cars is bound to increase in the future.

A petrol pump dispenses useful energy at a rate of 34 MJ s^{-1} = 34 MW

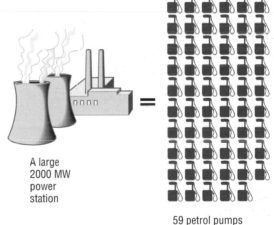

A large 2000 MW power station

59 petrol pumps

▲ **Figure 2** Just 59 petrol pumps can match the power output of a large power station.

What are the problems with petrol and diesel? One problem with petrol and diesel is that they are *finite* resources. Because crude oil supplies are limited, they probably won't last more than about another 100 years. This lifetime may extend as technology advances.

Also, we need crude oil for more than petrol and diesel alone. Crude oil provides the starting materials, or **feedstocks**, for the petrochemical industry. For example, it is needed for making synthetic fibres, detergents and pharmaceuticals. As we use up precious supplies, oil products may become too valuable to burn in car engines.

Another problem is pollution. Both petrol and diesel produce carbon dioxide when they burn, and that is a major contributor to the greenhouse effect (see **The Atmosphere** module, **Section A6**). Petrol and diesel also produce other kinds of emissions, as you'll see later in this module.

There is a real need to improve the performance of car engines so that they burn petrol and diesel as cleanly and efficiently as possible. There is also a need to find suitable fuels to replace petrol and diesel in the future. How chemists and chemistry can help is one of the major themes of this module.

But first we need to look at the chemistry behind combustion and ask where the energy comes from when a fuel burns.

DF2 *Getting energy from fuels*

Start by carrying out **Activity DF2.1** in which you compare the energy given out by burning hexane and methanol, both of which are found in some types of petrol. You will need some of the ideas in **Chemical Ideas 1.3** to do the calculations.

To answer questions like: 'How much energy can we get from a fuel?' and 'Which fuels store the most energy?' you have to know about 'thermochemistry'.

Thermochemistry is described in **Chemical Ideas 4.1** and you should read that now and look at some of the problems at the end of the section.

Activity DF2.2 shows you how chemists measure **enthalpy changes** that occur in solution and **Activity DF2.3** will help you to understand **Hess' law**.

The role of oxygen

Even the basic experimental method in **Activity DF2.1** shows that different fuels have different **enthalpy changes of combustion**. Let's compare six important fuels – look at Figure 3.

Note that the enthalpy changes are *negative* – this is because during combustion the reactants *lose* energy to their surroundings.

You can see that enthalpy changes of combustion vary widely. Why should this be? What decides how much energy you get when you burn a mole of a particular fuel?

We think of fuels as energy sources, but they can't release any energy until they have combined with

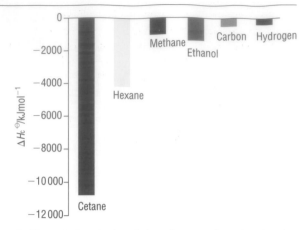

▲ **Figure 3** Standard enthalpy changes of combustion for some important fuels.

oxygen. So really we should think of *fuel–oxygen systems* as the energy sources. Burning fuels involves making and breaking chemical bonds.

Chemical Ideas 4.2 will tell you about making and breaking bonds. If you are uncertain about writing and balancing chemical equations, **Chemical Ideas 1.2** will remind you how to do this.

Assignment 1

a Write down balanced equations for the complete combustion of methane (CH_4), hexane (C_6H_{14}) and methanol (CH_3OH).

b Use the ideas of bond making and bond breaking to explain the following. (Think in terms of the bonds that have to be broken and the new bonds that are made when the fuel burns in oxygen, i.e. $\Delta H_c^{\ominus}$.)

   i Why is $\Delta H_c^{\ominus}$ for methane so much less negative than $\Delta H_c^{\ominus}$ for hexane?

   ii Why is $\Delta H_c^{\ominus}$ for methanol less negative than $\Delta H_c^{\ominus}$ for methane? After all, they both have the same number of C and H atoms.

The enthalpy change of combustion of a fuel depends on two things. First, there is the *number* of bonds to be broken and made – and that depends on the size of the molecule involved. That's why larger molecules such as hexane have a more negative $\Delta H_c^{\ominus}$ than smaller ones such as methane.

But $\Delta H_c^{\ominus}$ also depends on the *type* of bonds involved. Let's take your answer to b(ii) in Assignment 1 a bit further. The equations on page 23 show the bonds involved in the burning of methane and methanol.

The products are the same, but the key difference is that *methanol already has an O–H bond*. In other words, one of the bonds to oxygen is already made, unlike with methane where all the new bonds have to be made. Another way of looking at this is to say that methanol is methane which is already partly oxidised.

The energy released during combustion comes from the formation of bonds to oxygen. If methanol already has one bond made then it will give out less energy when it burns.

(a) Combustion of methane

$$CH_4 + 2O_2 \longrightarrow CO_2 + 2H_2O$$

$$
\begin{array}{c}
H \\
| \\
H-C-H \\
| \\
H
\end{array}
+
\begin{array}{c}
O=O \\
O=O
\end{array}
\longrightarrow
\quad O=C=O \quad +
\begin{array}{c}
H-O-H \\
H-O-H
\end{array}
$$

(b) Combustion of methanol

$$CH_3OH + 1\tfrac{1}{2}O_2 \longrightarrow CO_2 + 2H_2O$$

$$
\begin{array}{c}
H \\
| \\
H-C-H \\
| \\
O \\
| \\
H
\end{array}
\quad
\begin{array}{c}
O=O \\
\tfrac{1}{2}[O=O]
\end{array}
\longrightarrow
\quad O=C=O \quad +
\begin{array}{c}
H-O-H \\
H-O-H
\end{array}
$$

As a general rule, the more oxygen a fuel has in its molecule, the less energy it will give out when 1 mole of it burns. Oxygenated fuels such as alcohols and ethers are less energy-rich than hydrocarbon fuels. However, that's not to say that they are *poor* fuels. In some ways oxygenated fuels are better because they are less polluting than hydrocarbons, and they often have a high octane number. More about all this later.

Important news for slimmers

When you eat too much of an energy food, the excess energy gets stored in your body as fat. The more energy-rich the food, the more fattening it is.

Compare a carbohydrate, such as glucose, with a fat, such as olive oil. Here are the formulae:

glucose $\qquad\qquad$ $C_6H_{12}O_6$
glyceryl trioleate $\qquad$ $C_{57}H_{104}O_6$
(the main constituent of olive oil)

Per carbon atom, glucose is much more oxygenated than olive oil, so it is much *less* energy-rich. From 1 gram of a carbohydrate such as glucose you can get about 17 kJ. From 1 gram of a fat such as olive oil you can get about 39 kJ. Gram for gram, fats are more than twice as fattening as carbohydrates.

Alcohol is neither a fat nor a carbohydrate – but it too is bad news for slimmers. In fact there is a whole series of related compounds called **alcohols**, and the

particular alcohol present in drinks is ethanol, C_2H_5OH. The same substance is used as an alternative to petrol for cars in some countries. It burns in the car engine releasing energy – and it also releases energy when metabolised in the body (Figure 5).

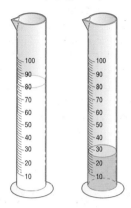

▲ **Figure 4** Fats and oils are more energy-rich than carbohydrates – the oil on the right will provide the same quantity of energy as the solid glucose on the left.

| 1 single measure of spirits | 1 glass of wine | $\tfrac{1}{2}$ pint of beer or lager |

▲ **Figure 5** Alcohol can be fattening: each of these drinks provides about 300 kJ of energy (about 70 Calories), equivalent to 1½ slices of bread.

Be thankful that you are not a potato! Humans store energy in fat but potatoes store it in starch. To store a given amount of energy, you need to have more than twice the mass of starch compared with fat. Think how oversize *that* would make you!

▲ **Figure 6** Why are these foods considered to be less fattening?

Table 1 Energy densities of some important fuels.

Fuel	Formula	Standard enthalpy change of combustion, $\Delta H_c^{\ominus}$/kJ mol^{-1}	Relative molecular mass	Energy density – energy transferred on burning 1 kg of fuel/kJ kg^{-1}
hexadecane (cetane)	$C_{16}H_{34}$	−10 700	226	−47 300
hexane	$C_6H_{14}(l)$	−4163	86	−48 400
methane	$CH_4(g)$	−890	16	−55 600
ethanol	$C_2H_5OH(l)$	−1367	46	−29 700
carbon	$C(s)$	−393	12	−32 800
hydrogen	$H_2(g)$	−286	2	−143 000

Carrying fuels around

The enthalpy change of combustion may not be the most important thing to consider for a practical fuel. What really matters is the *energy density* – how much energy you get per kilogram of fuel. After all, you have to carry the stuff around with you. We can work this out from the enthalpy change of combustion using the relative molecular mass. We've done this for six fuels and the results are in Table 1.

It is important that you make notes as you work through each section. **Activity DF2.4** will help you to check that you understand the main points in **Sections DF1** and **DF2**.

Assignment 2

Look at the values for the energy densities of the different fuels in Table 1.

a On the basis of energy density, which is the best fuel in the table? What are the practical difficulties involved in using this particular fuel?

b Compare hydrogen and hexane. Explain why hydrogen has the higher energy density, even though it has the lower (least negative) enthalpy change of combustion.

c Here are some data for octane, C_8H_{18}, and decane, $C_{10}H_{22}$, both of which are components of petrol:

	$\Delta H_c^{\ominus}$/kJ mol^{-1}	Relative molecular mass
octane, C_8H_{18}	−5470	114
decane, $C_{10}H_{22}$	−6778	142

Use the data to calculate the energy density for each of these compounds.

Compare your two answers. How do they compare with the values of energy density given for the fuels in Table 1?

▲ **Figure 7** The energy density of coal is relatively low, but when burnt it releases energy over an extended period, making it suitable for use in furnaces.

Later in this module we will come back to some of the fuels that might replace our current ones. But for now, let's look more closely at petrol and diesel.

DF3 *Focus on petrol and diesel*

'Sorry I'm late, the car wouldn't start …'

What do you blame if the car won't start on a cold morning? Almost certainly the car itself – probably not the driver, and certainly not the petrol. Yet using the correct petrol is actually very important.

What are petrol and diesel?

Petrol and diesel are both complex mixtures of many different compounds, carefully blended to give the right properties. The compounds are obtained from **crude oil** in several ways.

Crude oil is a mixture of many hundreds of **hydrocarbons**. It is a thick black liquid ('black gold' as it used to be called) but dissolved in it are gases and solids. Oil from the North Sea is pumped along pipes on the seabed to UK refineries, and special tankers bring crude oil from distant oilfields, such as those in the Middle East and Alaska. These refineries are either close to the shore (such as at Fawley, near Southampton) or the oil is off-loaded into a pipeline leading to a refinery (such as from Finnart on the west

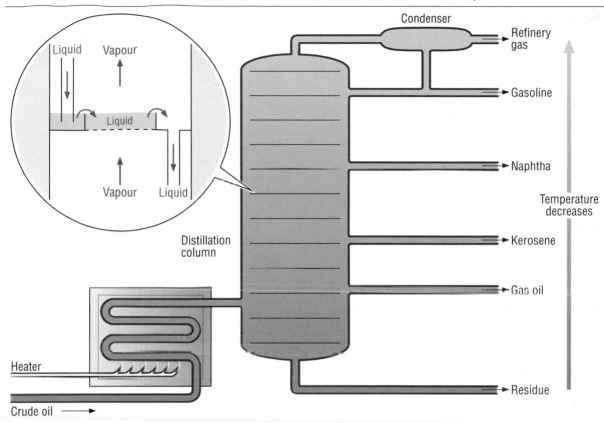

▲ **Figure 8** The primary fractional distillation of crude oil is a continuous process. Vapour rises up through the column and liquids condense and are run off at different levels, depending on their volatility.

coast of Scotland to the refinery at Grangemouth, near Edinburgh).

At the refinery the crude oil is heated to vaporise it and the vapour passes into a distillation column. In the column there is a temperature gradient – coolest at the top; hottest at the bottom. There are trays across the column with holes through which the rising vapour passes. The less volatile hydrocarbons condense on the trays and the more volatile ones pass through. This process is known as **fractional distillation** (Figure 8).

The oil is separated into fractions, each having a specific boiling point *range*. The fractions do not have an exact boiling point because they are *mixtures* of many different hydrocarbons. For example, the

gasoline fraction is a mixture of liquids, mostly **alkanes** with between five and seven carbon atoms, boiling in the range 25–75 °C.

The gasoline and *gas oil* fractions are the sources of petrol components. Another important fraction, *naphtha*, is also converted into high-grade petrol as well as being used in the manufacture of many organic chemicals. You can find information about the other fractions produced in the distillation of crude oil in Figure 8 and Table 2.

You can find out more about hydrocarbons and alkanes in **Chemical Ideas 12.1**.

Table 2 Fractions obtained from the fractional distillation of crude oil.

Name of fraction	Boiling point range/°C	Composition	% of crude oil	Use
refinery gas	<25	C_1–C_4	1–2	liquid petroleum gas (propane, butane), blending in petrol, feedstock for organic chemicals
gasoline	25–75	C_5–C_7		**car petrol**
naphtha	75–190	C_6–C_{10}	20–40	production of organic chemicals, converted to petrol
kerosene	190–250	C_{10}–C_{16}	10–15	jet fuel, heating fuel (paraffin)
gas oil	250–350	C_{14}–C_{20}	15–20	**diesel fuel**, central heating fuel, converted to petrol
residue	>350	>C_{20}	40–50	fuel oil (e.g. power stations, ships), lubricating oils and waxes, bitumen or asphalt for roads and roofing

There are two problems. The first is that the '*straight-run*' gasoline from the primary distillation makes poor petrol. Some is used directly in petrol but most is treated further. The second is a problem of supply and demand. Crude oil contains a surplus of the high boiling fractions, such as the gas oil and the residue, and not enough of the lower boiling fractions, such as gasoline. Although demand for gas oil itself is comparatively low, it can be cracked and used in car petrol, as you will see in **Section DF4**, therefore increasing demand. Figure 9 compares the supply and demand for different fractions of crude oil. You will see that demand is greater than supply for both petrol and diesel.

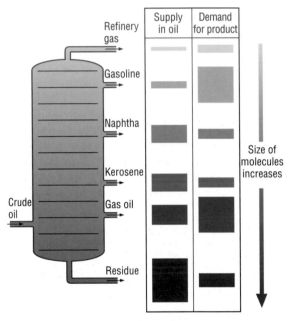

▲ **Figure 9** Supply and demand for different fractions of crude oil.

The job of the refinery is to convert crude oil into useful components. In order to do this, the structure of the alkane molecules present must be altered to produce different alkanes. The alkanes are also converted into other types of hydrocarbon that are used in petrol. These include **cycloalkanes**, which contain rings of carbon atoms, and **arenes** (**aromatic hydrocarbons**), which contain a benzene ring. The processes involved (isomerisation, reforming and cracking) are discussed later in this module. The products are blended to produce high-grade petrol.

Other products formed in refinery processes (for example, in cracking) include the **alkenes**, which contain a C=C double bond. Alkenes are needed to make some very useful compounds – for example, plastics, medicines and dyes.

You can look up the structures of alkenes and arenes in **Chemical Ideas 12.2** and **12.3**. You will study these sections in detail later. For now, you just need to be able to recognise these compounds from their structures.

The residue in the fractional distillation process can also be used to make useful products. First, it is distilled again, this time under reduced pressure in *vacuum distillation*. This avoids the high temperatures that would be needed for distillation at atmospheric pressure – such temperatures would tend to crack the hydrocarbons. The more volatile oil distils over, leaving behind a tarry residue. The oils are used as fuel oils in power stations or ships; some fractions are used as the 'base' oil in lubricating oils.

VACUUM DISTILLATION

This is a technique for distilling liquids which decompose when heated to their boiling point. Distillation cannot be carried out at atmospheric pressure, but if the external pressure is reduced then so is the boiling point, and distillation can be achieved below the decomposition temperature.

Winter and summer petrol

About 30–40% of each barrel of crude oil goes to make petrol. But you can see that it's not as simple as just distilling off the right bit at the refinery and sending it to the petrol stations. Petrol has to be blended to get the right properties. One important property is its *volatility*.

A mixture of petrol vapour and air is ignited in a cylinder in a car engine. The vapour–air mixture is made in the carburettor (see Figure 10). When the

Assignment 3

a   Which fraction from the primary distillation of crude oil would be most likely to contain each of the following hydrocarbons?
  i $C_{35}H_{72}$
  ii C_4H_{10}
  iii $C_{20}H_{42}$
  iv C_8H_{18}
b   What type of hydrocarbon are all of the above? Explain your answer.
c   The following hydrocarbons are all found in petrol. In each case, draw out one possible structural formula and state the type of hydrocarbon.
  i C_7H_8
  ii C_4H_8
  iii C_5H_{12}

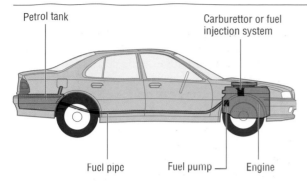

▲ **Figure 10** The fuel supply system. When you start a car engine, the fuel pump sucks petrol out of the petrol tank and pushes it to the carburettor. The carburettor partially vaporises the petrol, mixes it with air and sends it to the cylinders in the car engine (nowadays the carburettor is often replaced by an electronic fuel injection system).

weather is very cold the petrol is difficult to vaporise, so the car is difficult to start.

To get over this problem, petrol companies make different *blends* for different times of the year. During winter they put more volatile components in the petrol so it vaporises more readily. This means putting in more of the hydrocarbons with small molecules, such as butane and pentane.

On the other hand, in hot weather you don't want too many of these more volatile components, or the petrol will vaporise too easily. For one thing you'd lose petrol from your tank by evaporation – a process which is costly and polluting. Also, if the fuel vaporises too readily then pockets of vapour form in the fuel supply system. The fuel pump then delivers a mixture of liquid and vapour to the carburettor instead of mainly liquid. This means that not enough fuel gets through to keep the engine running – it's called *vapour lock*.

All petrols are a blend of hydrocarbons of high, medium and low volatility. As well as altering the petrol blend for the different seasons in a particular country, the blend will be different in different countries depending on the climate. The colder the climate, then more volatile components are added to the blend. Petrol companies change their blend four times a year – and you don't even notice. But you'd notice if they didn't!

You can compare the volatility of winter and summer blends of petrol in **Activity DF3.1**.

The problem of knocking

Another important property that blenders must take into account is the **octane number** of the petrol (see box on page 28). This is a measure of the tendency of the petrol to cause a problem known as '**knock**'.

▲ **Figure 11** Winter and summer petrol is not a new idea, as this advertisement shows.

Assignment 4

Petrol blenders talk about the 'front-end' (high volatility), 'mid-range' and 'tail-end' (low volatility) components of a blend.

a   What differences will there be between
   i   a winter and a spring blend for the UK?
   ii   a summer blend for Spain and a summer blend for the Netherlands?
b   Use the **Data Sheets** to look up the densities of some alkanes.
   i   How does the density of alkanes change as their molecular mass increases?
   ii   Which will have the greater mass at a given temperature – 1 litre of petrol bought in the Netherlands or 1 litre bought in Spain? Explain your answer.
c   'Spanish people get a better bargain than Dutch people when they buy petrol.' Discuss this statement.

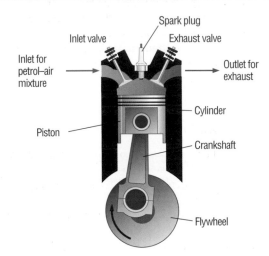

▲ **Figure 12** How a four-stroke petrol engine works. The compression stroke is shown here. The piston compresses the petrol–air mixture, then a spark makes the mixture explode, pushing the piston down and turning the crankshaft. In a diesel engine, air alone enters via the inlet valve. There is no spark plug and fuel is injected via a separate inlet during the compression stroke. The high pressure causes the fuel–air mixture to ignite.

In both petrol and diesel engines, the fuel–air mixture has to ignite at the right time, usually just before the piston reaches the top of the cylinder.

Look at Figure 12. As the fuel–air mixture is compressed it heats up, and the more it is compressed the hotter it gets. Modern cars achieve greater efficiency than in the past by using *higher compression ratios*, often compressing the gases in the cylinder by about a factor of 10.

Many hydrocarbons **auto-ignite** under these conditions. The fuel–air mixture catches fire as it is compressed. When this happens, *two* explosions occur:

one due to the compression and another when the spark occurs. This produces a '**knocking**' or 'pinking' sound in the engine. The thrust from the expanding gases is no longer occurring at the proper time, so engine performance is lowered and the inside of the combustion cylinder can be damaged.

The auto-ignition of hydrocarbons is explored in **Activity DF3.2**.

OCTANE NUMBERS

The tendency of a petrol-based fuel to auto-ignite is measured by its **octane number**. 2,2,4-trimethylpentane (which used to be called 'iso-octane' – hence the name of the scale) is a branched alkane with a low tendency to auto-ignite. It is given an octane number of 100. Heptane, a straight-chain alkane, auto-ignites easily and is given an octane number of 0.

The octane number of any fuel is the percentage of 2,2,4-trimethylpentane in a mixture of 2,2,4-trimethylpentane and heptane which knocks at the same compression ratio as the given fuel. For example, four-star petrol has an octane number of 97 and knocks at the same compression ratio as a mixture of 97% 2,2,4-trimethylpentane and 3% heptane.

$$CH_3-\underset{\underset{CH_3}{|}}{\overset{\overset{CH_3}{|}}{C}}-CH_2-\overset{\overset{CH_3}{|}}{CH}-CH_3$$

2,2,4-trimethylpentane   low tendency for auto-ignition scores 100

$$CH_3-CH_2-CH_2-CH_2-CH_2-CH_2-CH_3$$

heptane   high tendency for auto-ignition scores 0

Diesel

Diesel fuel is a mixture of much larger hydrocarbon molecules than those found in petrol. Diesel that comes from crude oil contains about 75% of saturated hydrocarbons (mostly alkanes and **cycloalkanes**) and 25% of **aromatic hydrocarbons** (including naphthalenes and alkylbenzenes). An average chemical formula for ordinary diesel fuel is $C_{12}H_{26}$, with a range from $C_{10}H_{22}$ to $C_{15}H_{32}$.

As previously stated, a petrol engine takes in a mixture of vapour and air, compresses it and then ignites the mixture with a spark. A diesel engine takes in just air, compresses it and then injects fuel into the compressed air (see Figure 12). The heat of the

Assignment 5

In this assignment you will use data on the octane number of different alkanes to see whether there is any relationship between the structure of an alkane and its octane number. The assignment is also a useful opportunity to practise naming alkanes.

Table 3 The octane number of different alkanes

Name of alkane	Octane number
heptane	0
hexane	25
pentane	62
3-methylhexane	65
2-methylpentane	73
3-methylpentane	75
2,3-dimethylpentane	91
2-methylbutane	93
butane	94
2-methylpropane	>100

a For each of the alkanes named in the table, write the full structural formula.

b Which of the alkanes are isomers of one another? Sort them into groups of isomeric alkanes.

c Classify each alkane as:
  S   straight-chain
  B   branched-chain
  M   multiple-branched chain.

d Plot a graph showing the octane number of the alkane against the number of carbon atoms in the molecule. Label each point on the graph with the type of alkane concerned (S, B or M).

e What do you conclude about:
  i the effect of chain length on octane number?
  ii the effect of chain branching on octane number?

f What is the significance of all this for petrol blending?

compressed air ignites the fuel spontaneously. Thus a diesel engine works by auto-ignition, something that must be avoided in a petrol engine!

DF4 *Making petrol – getting the right octane number*

Different cars have different compression ratios. High-performance petrol engines usually have a high compression ratio, and they need a high octane fuel –

otherwise there would be knocking and the engine would suffer. There are two ways of dealing with the knocking problem. One is to put special additives in the petrol which discourage auto-ignition. The other is to blend high-octane compounds with the ordinary petrol.

Look – no lead

Anti-knock additives are substances which reduce the tendency of alkanes to auto-ignite. They increase the octane number of the petrol.

From the 1920s until 1992, petrol engines were designed to run on petrol with small amounts of lead compounds added. These were used as economical and effective anti-knock additives. Exactly how they work is still not fully understood but they helped to prevent the reactions which cause knocking.

However, concern over environmental effects led to a gradual phasing-out of leaded petrol. For one thing, the lead compounds present in the exhaust fumes are toxic. They also poison the metal catalysts in the catalytic converters installed to reduce the levels of other pollutants in exhaust fumes.

Rather than develop alternative lead-free additives, the petrol companies turned their attention to using refining and blending to get high octane numbers without using lead.

Refining and blending

The hydrocarbons which give the best performance in a petrol engine are not the ones which are most plentiful in crude oil. So it is the job of the refinery to 'doctor' the hydrocarbons to suit our needs.

Mostly, it is a case of getting the right kind of alkane – although other types of hydrocarbons and some oxygenated compounds are also important, as you'll see later.

Which alkanes?
The structure of an alkane has an important influence on its tendency to auto-ignite – in other words, on its octane number. In general, the shorter the alkane chain then the higher the octane number. Short-chain alkanes are also more volatile, of course, so they can be used both to increase the octane number and to improve cold-starting. Even gaseous alkanes such as butane can be used – they just dissolve in the petrol.

The idea of isomerism is important in this section. You can find out about it in **Chemical Ideas 3.3**.

Petrol blenders are restricted in the proportion of short-chain alkanes they can include in the blend – too much and the petrol blend would be too volatile.

The other factor that affects octane number is the *degree of branching* in the alkane chain. Quite simply, the more branched the chain then the higher the octane number.

Crude oil contains both straight-chain and branched alkanes. Unfortunately, it does not contain enough of the branched isomers to give it a naturally high octane number. The octane number of 'straight-run' gasoline is about 70.

To get around this the petrol companies have a number of clever ways of increasing the octane number of petrol. These include **isomerisation**, **reforming** and **cracking**.

You can make models of alkane isomers and practise naming them in **Activity DF4.1**.

In **Activity DF4.2** you can practise drawing alkane structures using a computer package.

Isomerisation

Isomerisation involves taking straight-chain alkanes, heating them in the presence of a suitable catalyst so that the chains break, and then letting the fragments join together again. When the fragments join they are more likely to do so as *branched* chains rather than as *straight* chains.

Oil refineries do this on a large scale with pentane (C_5H_{12}) and hexane (C_6H_{14}), both products of the distillation of crude oil. It is an important way of increasing the octane number of petrol.

One isomerisation reaction which can happen with pentane is shown below:

$$CH_3-CH_2-CH_2-CH_2-CH_3$$

pentane
octane number 62

$$\rightleftharpoons$$

$$CH_3-\underset{\underset{\displaystyle CH_3}{|}}{CH}-CH_2-CH_3$$

2-methylbutane
octane number 93

Isomerisation reactions like this do not go to completion and a mixture is formed containing all the possible isomers. We say the reaction has come to a **state of equilibrium** when no further change is possible under the reaction conditions. The arrows $\rightleftharpoons$ indicate that the reaction is reversible and forms an equilibrium mixture. (You will study equilibrium reactions in detail later in the course.)

In a modern plant, the isomerisation takes place in the presence of a catalyst of aluminium oxide on which platinum is finely dispersed. The products then pass over a form of **zeolite**, which acts as a *molecular sieve* and separates the straight-chain from the branched isomers. The straight-chain alkanes are then recycled

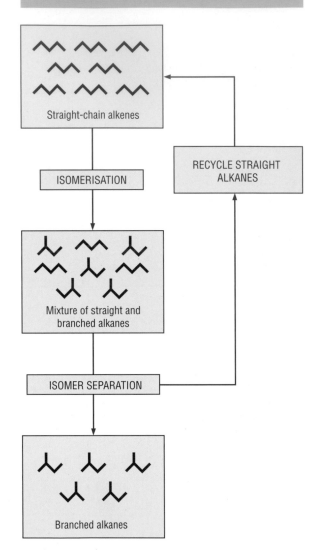

▲ **Figure 13** The isomerisation process.

over the platinum catalyst. Figure 13 gives a flow diagram for the process.

Reforming

Reforming is another technique used by oil refiners to increase the octane quality of petrol components. Naphtha, which is mainly made up of straight-chain alkanes with 6–10 carbon atoms, is heated to about 500 °C and passed over a catalyst. The straight-chain

ZEOLITES

Zeolites belong to a large family of naturally occurring minerals containing mainly aluminium, silicon and oxygen. They can also be made synthetically.

There are many different zeolites because of the different ways the atoms can be arranged in the crystal structure. They all contain an extensive network of interlocking pores and channels, which provide a large surface area. These pores and channels are of different sizes in different zeolites.

Zeolites are used widely in industry as **catalysts**, and also as *molecular sieves* to sort out molecules by size and shape (Figure 14).

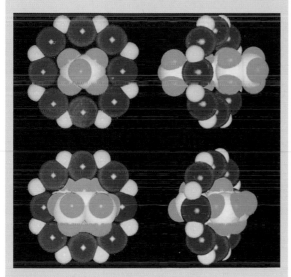

▲ **Figure 14** This computer-generated picture shows a molecular sieve in action. In the top pictures, a straight-chain molecule of hexane passes through a molecular sieve. In the bottom pictures, a branched-chain molecule of dimethylbutane cannot pass through. Zeolite molecular sieves are used to separate straight-chain and branched-chain isomers in the isomerisation process.

alkanes are converted to ring compounds – first to cycloalkanes and then to aromatic hydrocarbons. Some typical reforming reactions are shown in Figure 15.

The catalyst is platinum, which is finely dispersed on the surface of aluminium oxide. The process is called **platforming**. During the process some of the hydrocarbons decompose to carbon, which decreases the efficiency of the catalyst. There can be £5 million worth of platinum inside a single platformer, so it is very important to regenerate the catalyst and keep it in prime working order. Excess hydrogen is mixed with the naphtha going into the process to suppress the formation of carbon.

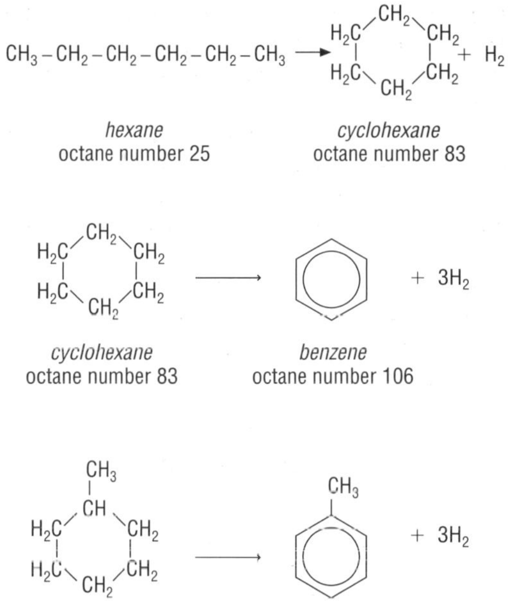

$$CH_3-CH_2-CH_2-CH_2-CH_2-CH_3 \longrightarrow \text{cyclohexane} + H_2$$

hexane
octane number 25

cyclohexane
octane number 83

cyclohexane
octane number 83

$\longrightarrow$ benzene $+ 3H_2$

benzene
octane number 106

methylcyclohexane
octane number 70

$\longrightarrow$ methylbenzene $+ 3H_2$

methylbenzene
octane number 120

▲ **Figure 15** Some typical reforming reactions showing octane numbers of reactions and products.

Cracking: using the whole barrel

Cracking is one of the most important reactions in the petroleum industry. It starts with alkanes that have large molecules that are too big to use in petrol – for example, alkanes from the gas oil fraction. These large molecules are broken down to give alkanes with shorter chains that can be used in petrol. What's more, these shorter-chain alkanes tend to be highly branched, so petrol made by cracking has a higher octane number.

There's another benefit from cracking: it also helps to solve the supply and demand problem (see Figure 9, page 26).

Cracking: how is it done?

Much of the cracking carried out to produce petrol is done by heating heavy oils, such as gas oil, in the presence of a catalyst. It is called catalytic cracking or 'cat cracking' for short. The molecules in the feedstock can have 25–100 carbon atoms, although most will usually have 30–40 carbon atoms.

Cracking reactions are quite varied. Some of the types of reactions are:

*   alkanes → branched alkanes + branched alkenes;
    An example of this is:

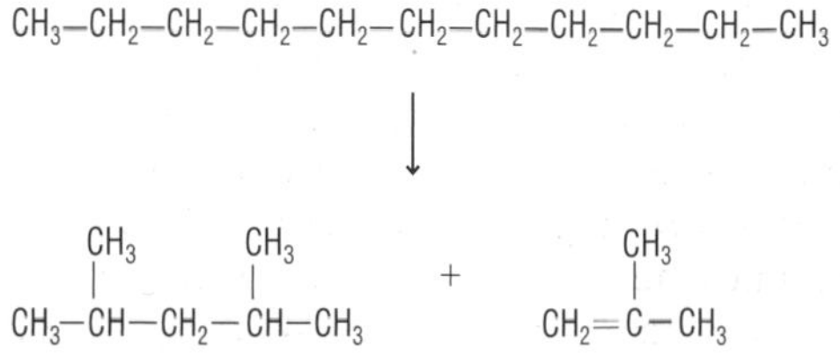

$$CH_3-CH_2-CH_2-CH_2-CH_2-CH_2-CH_2-CH_2-CH_2-CH_2-CH_3$$

$$\downarrow$$

$$CH_3-\underset{\underset{CH_3}{|}}{CH}-CH_2-\underset{\underset{CH_3}{|}}{CH}-CH_3 \quad + \quad CH_2=\underset{\underset{CH_3}{|}}{C}-CH_3$$

- alkanes → smaller alkanes + cycloalkanes;
- cycloalkanes → alkenes + branched alkenes;
- alkenes → smaller alkenes.

The alkenes which are produced are important feedstocks for other parts of the petrochemicals industry.

Cracking always produces many different products, which need to be separated in a fractionating column.

Once again, zeolites play an important role but this time as catalysts. Type Y zeolite is particularly effective in producing good yields of high octane number products.

In a modern cat cracker, the cracking takes place in a 60 m high vertical tube about 2 m in diameter (Figures 16 and 17). It is called a riser reactor because the hot vaporised hydrocarbons and zeolite catalyst are fed into the bottom of the tube and forced upwards by steam. The mixture is a seething fluidised bed in which the solid particles flow like a liquid.

It takes the mixture about 2 seconds to flow from the bottom to the top of the tube – so the hydrocarbons are in contact with the catalyst for a very short period of time.

One of the problems with cat cracking is that, in addition to all the reactions you have already met, coke (carbon from the decomposition of hydrocarbon molecules) forms on the catalyst surface so that the catalyst eventually becomes inactive. The powdery catalyst needs to be *regenerated* to overcome this problem.

After the riser reactor, the mixture passes into a separator where steam carries away the cracked products leaving behind the solid catalyst. This then goes into the regenerator, where it takes about 10 minutes for the hot coke to burn off in the stream of air which is blown through the regenerator. The catalyst is then reintroduced into the base of the reactor ready to repeat the cycle.

The energy released from the burning coke heats up the catalyst. The catalyst transfers the energy to the feedstock so that cracking can occur without additional heating.

You can find out more about cracking, and try cracking alkanes for yourself, in **Activities DF4.3** and **DF4.4**.

Cat crackers have been in operation since the late 1940s and have become very flexible and adaptable. They can handle a wide range of different feedstocks. The conditions and catalyst can be varied to give the maximum amount of the desired product – in this case branched alkanes for blending in petrol.

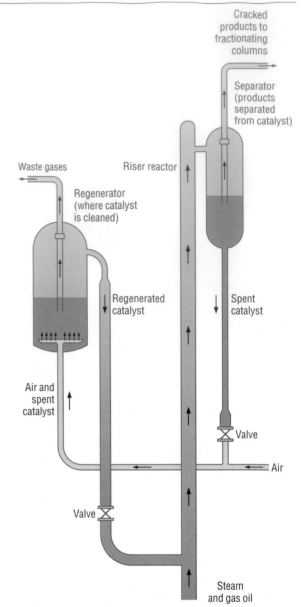

▲ **Figure 16** How a cat cracker works. The feedstock is gas oil and the cracking reaction takes place in the riser reactor.

Assignment 7

The riser reactor uses a fluidised bed of solid catalyst and reactants. Fluidised beds have the following useful properties.
- The solid can flow along pipes.
- As the solid can flow, it is easier to make the reaction into a continuous process.
- There is very good contact between the gas and the large surface area of the solid catalyst.
- Energy transfer between the solid and the gas is very efficient.

Suggest why each of these properties is useful or important in the cat cracker.

▲ **Figure 17** A cat cracker which converts gas oil into branched alkanes for blending to produce high octane petrol. The process is shown in detail in Figure 16.

Assignment 8

a Using molecular formulae, write out a balanced equation for one of the conversions shown in Figure 15.

b Explain why these reactions are not isomerisations.

c There is a list of alkanes below. For each one, say whether you think:

   A It could be used unchanged in petrol.

   B It should go through a reforming process before blending to form petrol.

   C It should be cracked before blending to form petrol.

     i $CH_3CH_2CH_2CH_2CH_2CH_3$

     ii $CH_3(CH_2)_{15}CH_3$

     iii $(CH_3)_3CCH(CH_3)CH_3$

     iv $CH_3CH_2CH_2CH_3$

Adding oxygenates

Oxygenates is the name the petrol blenders use for fuels containing oxygen in their molecules. Two types of compounds are commonly used: **alcohols** and **ethers** (see box).

ALCOHOLS AND ETHERS

Alcohols all have an OH group in their molecule. The best known alcohol is ethanol, commonly called just 'alcohol':

$$\begin{array}{c} \quad H \quad H \\ \quad | \quad \; | \\ H-C-C-O-H \\ \quad | \quad \; | \\ \quad H \quad H \end{array}$$

Ethers all have an oxygen atom bonded to two carbon atoms. An example is ethoxyethane (sometimes just called 'ether'):

$$\begin{array}{c} \quad H \quad H \qquad H \quad H \\ \quad | \quad \; | \qquad \; | \quad \; | \\ H-C-C-O-C-C-H \\ \quad | \quad \; | \qquad \; | \quad \; | \\ \quad H \quad H \qquad H \quad H \end{array}$$

You can find out more about alcohols and ethers in **Chemical Ideas 13.2**.

You can practise naming alcohols in **Activities DF4.5** and **DF4.6**.

Table 4 shows the oxygenates most commonly used for blending with petrol, together with 'straight-run' gasoline for comparison.

An oxygenate that is commonly used is MTBE ('methyl tertiary butyl ether', the name which was once used for this compound; its modern systematic name is 2-methoxy-2-methylpropane).

$$\begin{array}{c} \qquad\quad CH_3 \\ \qquad\quad | \\ CH_3-C-O-CH_3 \qquad MTBE \\ \qquad\quad | \\ \qquad\quad CH_3 \end{array}$$

Figure 18 (page 34) shows the percentages of MTBE added to different grades of petrol.

Table 4 Oxygenates commonly used for blending with petrol ('straight-run' gasoline is given for comparison).

Name	Formula	Homologous series	Octane number	Boiling point/°C	Relative cost per litre (approx)
methanol	CH_3OH	alcohols	114	65	1.0
ethanol	CH_3CH_2OH	alcohols	111	79	5.8
MTBE	$CH_3OC(CH_3)_3$	ethers	113	55	1.6
'straight-run' gasoline	–	–	70	–	1.1

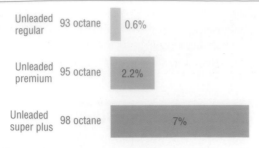

▲ **Figure 18** Percentages by volume of MTBE typically used in different blends of petrol.

Oxygenates are added to petrol for two reasons. Firstly they increase the octane number, as you can see from Table 4. Secondly, levels of carbon monoxide in the exhaust are reduced. However, these claims are not universally accepted. Moreover, some states in the US have banned the use of MTBE in petrol. It is soluble in water and spillages can pollute water supplies. In the United Kingdom, MTBE continues to be added to petrol but in lower amounts (<1%) than were previously used in America.

The perfect blend

Figure 19 shows the octane numbers of some of the key ingredients at the petrol blender's disposal. The blender's job is to produce petrol from these ingredients at minimum cost. Of course, the petrol must meet

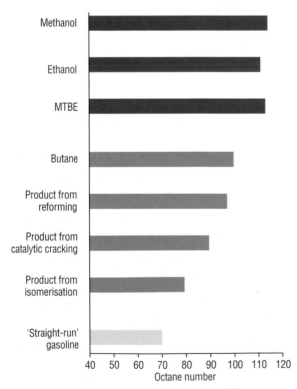

▲ **Figure 19** Octane numbers of some of the key ingredients used in petrol blending. Note the high octane numbers of the three oxygenates.

specifications about volatility, octane number, density, etc. Prices and availability fluctuate and refiners are helped in their decisions by computer models of their refinery. They must take into account not only the prices of crude oil and the final petrol, but prices of their other products from the refinery and energy costs too.

Blending is done in batches of around 20 000 000 litres at a time and can take as long as 20 hours to complete. Thorough mixing is important to give a homogeneous liquid.

DF5 *Making diesel fuel*

Whereas petrol blends are given octane numbers, diesel fuels have cetane numbers. However, the two terms are not equivalent. This is because petrol's desirable property is to resist auto-ignition to prevent engine knocking, whereas diesel's desirable property is to auto-ignite. Cetane is hexadecane, $C_{16}H_{34}$, which ignites very easily under compression. Hexadecane is given a **cetane number** of 100. An aromatic compound called methylnaphthalene has very low tendency to ignite and is given a cetane number of 0. All other hydrocarbons in diesel fuel are indexed to cetane/methylnaphthalene mixtures depending on how well they ignite under compression. The cetane number of the whole fuel measures how well the fuel auto-ignites in the diesel engine. A high cetane number means that there will be a high tendency to auto-ignite. Typical cetane numbers are around 50, though there is very little cetane in most diesel fuels.

Other important features of diesel fuel are:

- *Lubricity* – previously sulfur compounds in diesel provided this lubrication feature but, with these being removed (see below), other additives are now used.
- *Flash point* – this must be high for diesel, just as it is low for petrol, since diesel is not required to burn when ignited or sparked.
- *Cloud point* – being composed of much longer chain compounds, diesel is more likely to solidify in cold conditions and this results in the fuel looking cloudy. Summer and winter diesels are sold, with winter diesel having a lower cloud point.

Why do hydrocarbons mix?

The hydrocarbons in both petrol and diesel are completely miscible with each other and can be blended in any proportions. Have you ever wondered why some liquids mix easily and others, like oil and vinegar, don't mix at all? Mixing of two substances is a natural change – it happens by chance with no external help. The result is an increase in *disorder*. The idea of disorder, or randomness, in a system is very important in determining the direction in which changes occur.

Assignment 9

Table 5 The cetane number of different alkanes.

Name	Formula	Skeletal formula	Cetane number
tetradecane	$C_{14}H_{30}$		95
pentadecane	$C_{15}H_{32}$		98
hexadecane (cetane)	$C_{16}H_{34}$		100
heptadecane	$C_{17}H_{36}$		105
octadecane	$C_{18}H_{38}$		110
2,2,4,4,6,8,8-heptamethylnonane	$C_{16}H_{34}$		15
7,8-dimethyltetradecane	$C_{16}H_{34}$		40

a Copy Table 5 and insert the skeletal formulae of the compounds.
b Which of the compounds are structural isomers?
c Comment on the effect of chain length and chain branching on the cetane number of a hydrocarbon.
d How do the effects commented on in part **c** compare with the effects of chain length and branching on octane number?

We give the amount of disorder in a system a name – **entropy**. When the system gets more disordered, like a pack of cards being shuffled or two liquids mixing, its entropy increases.

All substances tend to mix with one another, thereby increasing the entropy of the system, unless there is something stopping them. In the case of oil and vinegar there are attractive forces between the particles in vinegar which prevent them mixing with the oil particles.

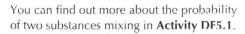

The idea of entropy is discussed in more detail in **Chemical Ideas 4.3**.

You can find out more about the probability of two substances mixing in **Activity DF5.1**.

Activity DF5.2 will help you check your knowledge and understanding of all work you have done on petrol so far.

DF6 *Trouble with emissions*

A serious problem

We'll look next at a problem of motor fuels which is causing worldwide concern: exhaust emissions. Figure 20 shows what goes into – and what comes out of – a car engine. There are slight differences between the emissions of petrol and diesel engines (which we will deal with later) but many of the pollutants are similar. The nitrogen oxides, NO and NO_2, are grouped together as NO_x, although the NO_x emitted by vehicle exhausts contains mainly NO with only small quantities of NO_2. Similarly, SO_x represents the oxides of sulfur, SO_2 and SO_3, and C_xH_y represents the various hydrocarbons present in the exhaust fumes.

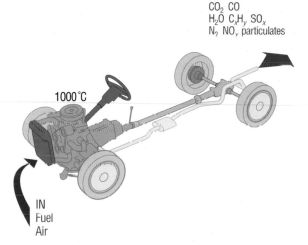

OUT
CO_2 CO
H_2O C_xH_y SO_x
N_2 NO_x particulates

1000°C

IN
Fuel
Air

▲ **Figure 20** What goes into – and comes out of – a car engine.

Another problem arises because petrol is so volatile. On a warm day a parked car gives off hydrocarbon fumes, mostly butane, from the petrol tank and the carburettor. This is called *evaporative emission* and accounts for about 10% of emissions of volatile organic compounds from petrol vehicles.

- The oxides of sulfur in vehicle exhausts come from sulfur compounds in the fuel. These combine with the oxygen in the air in the heat of the engine.
- Oxides of nitrogen are formed mainly from the components of the air itself. At the high temperatures of vehicles' engines, the nitrogen and the oxygen in the air react to form NO. Some of this reacts with more oxygen to form NO_2.
- Oxides of sulfur and nitrogen are acidic and give rise to acid rain in the atmosphere. This can cause health problems, particularly for asthmatics, corrode limestone buildings and damage forests and lakes.

- Carbon monoxide is formed by the incomplete combustion of hydrocarbon fuels. It is very toxic to humans – it is oxidised to carbon dioxide in the atmosphere.
- Particulates are very small carbon particles (not visible to the naked eye) which can get into our lungs and cause irritation. Particulates are also produced by incomplete combustion of the hydrocarbon fuels in diesel.

You can compare the acidity of some polluting gases in **Activity DF6.1**.

▲ **Figure 21** Views of Denver, Colorado, US.
a A computer-generated view showing what the visibility would be like without photochemical smog.
b A photograph showing the current smoggy conditions.

Photochemical smog

Unfortunately, the substances shown in Figure 20 are not the only pollutants caused by motor vehicles. **Ozone** is a *secondary pollutant* because it is not released directly into the atmosphere. It is formed as a result of chemical reactions that take place when sunlight shines on a mixture of two of the *primary pollutants*, nitrogen oxides (NO_x) and hydrocarbons (C_xH_y) together with oxygen and water vapour.

Also produced are irritating and eye-watering compounds formed by the breakdown and further reaction of the hydrocarbons. These reactions all occur in *photochemical smogs* which are a great cause for concern. (A photochemical reaction occurs when a molecule absorbs light energy and then undergoes a chemical reaction.)

OZONE – A MOLECULE OF MANY PARTS

Ozone is a highly reactive substance with molecules containing three oxygen atoms – its formula is O_3. Ozone occurs in the atmosphere, both high up in the stratosphere and closer to the ground in the troposphere. Its presence in these two regions affects us in very different ways.

In the stratosphere, ozone acts as a sunscreen filtering out dangerous ultraviolet light from the incoming solar radiation. Its presence there is vital to our survival, though there is now great concern because chlorofluorocarbons (CFCs) reaching the stratosphere are causing depletion of the protective ozone layer. You will find out more about this in **The Atmosphere**.

Meanwhile, in the troposphere ozone has an important role in the production of hydroxyl radicals (HO). These are short-lived species, formed from water, which bring about the breakdown of many substances that would otherwise build up to hazardous levels in the atmosphere. However, ozone is an irritant toxic gas and high concentrations near ground level are detrimental to human health. It weakens the body's immune system and attacks lung tissue. Also, ozone in the troposphere acts as a greenhouse gas and so contributes to global warming.

Photochemical smog contains a mixture of primary and secondary pollutants (see Figure 22). Its exact composition varies enormously and depends, for example, on the nature of the primary pollutants, the local geography, weather conditions, the time of day and the length of the smog episode.

Photochemical smogs normally occur in the summer during high pressure (anticyclonic) conditions. The still air means that there is much less mixing with high altitude air and the pollutants are trapped near ground level. Figure 23 summarises the formation of photochemical smog. Note that, even in clean relatively 'unpolluted' air, there is a small background concentration of ozone and this is involved in the series of reactions producing the photochemical

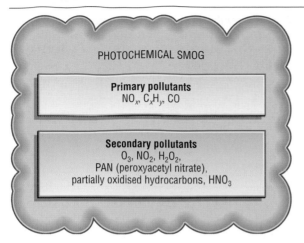

▲ **Figure 22** The main constituents of photochemical smog.

smog; its concentration is *increased* during smog formation.

If there is a light prevailing wind then the polluted air mass will be transported from the built-up urban area where it was generated and will move across rural areas. In fact, the highest ozone readings are often recorded over rural areas because the chemical reactions producing the secondary pollutants take place relatively slowly.

What are the effects of photochemical smog?

Photochemical smogs cause haziness and reduced visibility in the air close to the ground. For many people they can cause eye and nose irritation and some difficulty in breathing, but for vulnerable groups – such as asthmatics who already have respiratory problems, very young children and many old people – the effects can be more serious.

Ozone is not the only substance in smog to cause health problems. The polluted air contains a whole cocktail of harmful chemicals, but the exact links between photochemical smog and health problems such as asthma are very difficult to establish and are still the subject of much research. There is no simple relationship between the two because so many other factors are involved.

Humans are not the only ones to suffer. High ozone concentrations affect animals and plants too. Ozone is a highly reactive substance which attacks most organic matter. Compounds with carbon–carbon double bonds are particularly vulnerable so many materials, such as plastics, rubbers, textiles and paints, can be damaged.

How can chemists help?

Research work is taking place on several fronts to unravel and explain the complex chemistry involved.

Monitoring pollutants

An important first step is to find out exactly which pollutants are present in the troposphere and how their concentrations vary. For example, ozone concentrations in urban areas show regular fluctuations which reflect changes in sunlight and vehicle emissions during the day – see Figure 24 (page 38). There are now several networks of monitoring stations across the UK that record pollutant concentrations. Figure 25 (page 38) shows the location of ozone monitoring sites in the UK.

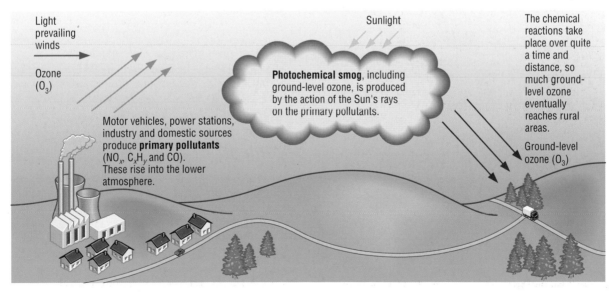

▲ **Figure 23** Formation of photochemical smog.

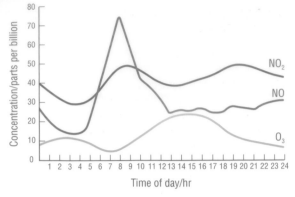

▲ **Figure 24** Variation of the concentrations of ozone and oxides of nitrogen on a summer weekday in London. The high NO peak coincides with the early morning traffic rush hour, and is quickly followed by a build-up of NO_2 as reactions in the air convert NO into NO_2. Photochemical reaction of NO_2 leads to formation of O_3 which reaches its peak in the early afternoon. The evening rush hour generates more NO and NO_2 but decreasing sunlight slows down photochemical reactions and there is no corresponding build-up of ozone.

Studying individual reactions in the laboratory

The air around us may look very still and unchanging – but this is deceptive. The troposphere acts as a huge reaction mixture with a vast array of interrelated chemical reactions taking place all the time. Many of these reactions involve 'broken down' fragments of molecules, called **radicals**, and the reactions can take place very quickly indeed. Other reactions take place more slowly. To make predictions about pollution, chemists need to know what reactions can take place

and how quickly they occur. This means investigating each reaction in the laboratory and measuring the rate at which it takes place under a variety of carefully controlled conditions.

Modelling studies

Much of the information on rates of reactions is used in computer-simulation studies which aim to reproduce and predict the behaviour of pollutants during a smog episode.

Smog chamber simulations

These are laboratory experiments on a grand scale. Primary pollutants are mixed in a huge plastic bag called a smog chamber and are then exposed to sunlight under carefully controlled conditions – see Figure 26. Analytical probes monitor the concentrations of various species as the photochemical smog builds up.

▲ **Figure 26** The EU smog chamber in Valencia, Spain. The reaction chamber has a volume of approximately 200 m³ (200 000 litres). It has to be this big to reduce to a minimum any 'surface effects' where reactions take place on the walls of the container instead of in the gas phase.

You can consolidate your understanding of vehicle pollutants by doing **Activity DF6.2**.

DF7 Tackling the emissions problem

Concern about air pollution from motor vehicles is mounting worldwide and many countries are bringing in legislation to limit emissions. The US has led the way and cars are put through a rigorous emissions test cycle before they are allowed on the road. Figure 27 shows how emission limits have become increasingly severe in Europe.

There are two ways of tackling the emissions problem directly. One involves changing the design of

▲ **Figure 25** Automatic ozone monitoring sites in the UK.

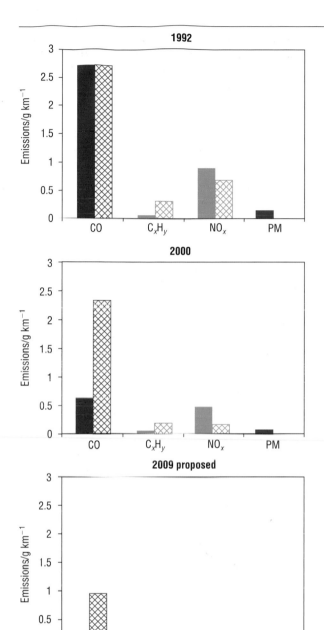

▲ **Figure 27** European emission limits for new vehicles are becoming increasingly severe ('PM' stands for 'particulate matter'.).

cars, and the other involves changing the fuel used by the car.

Of course, there are important indirect methods for tackling the emissions problem, such as limiting the traffic entering towns and encouraging car-sharing schemes for people travelling to and from work.

However, in this section we shall concentrate on methods of reducing the emissions from a single car by changing the technology of the car engine and by modifying the exhaust system.

Changing the engine technology of petrol engines

How much air does a petrol engine need?

Chemical Ideas 1.4 will help you to understand how to calculate volumes of gases and **Activity DF7** will give you some practice in dealing with gases and carrying out such calculations.

You can work out the amount of air needed by a petrol engine by doing Assignment 10.

Assignment 10

For the purposes of this assignment, assume that petrol is pure heptane, C_7H_{16}.
a   Write an equation for the complete combustion of heptane.
b   How many moles of oxygen are needed for the combustion of 1 mole of heptane?
c   How many moles of heptane are there in 1 g?
d   What mass of oxygen is needed to burn 1 g of heptane? (A_r: O = 16)
e   What mass of air is needed to burn 1 g of heptane? (Assume that air is 22% oxygen and 78% nitrogen by *mass*.)
f   What *volume* of air (measured at 25 °C and 1 atmosphere pressure) is needed to burn 1 g of heptane? (Assume that 1 mole of oxygen has a volume of 24 dm³ under these conditions and that air is 21% oxygen and 79% nitrogen by *volume*.)

The ratio of air to fuel which you worked out in Assignment 10 is the ratio needed to make petrol burn completely. It is called the *stoichiometric ratio*. (*Stoichiometric* means 'involving the exact amounts shown in the chemical equation'.)

For an average car engine, the stoichiometric ratio is usually taken to be about 15:1 (by mass).

If you have *less* air than that in the stoichiometric ratio, it is called a 'rich' mixture – because it is rich in petrol. A 'lean' mixture has *more* air than in the stoichiometric ratio – it produces less NO_x and CO but the levels of C_xH_y can actually increase (see Figure 28). The trouble is that if the mixture is *too* lean then the engine misfires and emissions increase.

'Lean-burn' engines use an air:fuel ratio of around 18:1 and have specially designed combustion chambers to get over the problem of misfiring.

Another advantage of a lean-burn engine is that it gives better fuel economy because you are using less fuel for each firing of the cylinder.

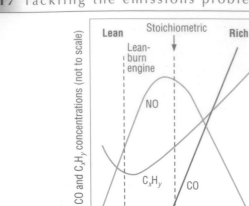

▲ **Figure 28** How gases in exhaust emissions vary with the composition of the air–fuel mixture.

Using catalysts

Catalysts speed up chemical reactions without getting used up themselves. We can use them to speed up reactions which involve pollutants in car exhausts. Look at these reactions:

$$2CO(g) + O_2(g) \rightarrow 2CO_2(g) \qquad \text{(reaction 1)}$$
$$C_7H_{16}(g) + 11O_2(g) \rightarrow 7CO_2(g) + 8H_2O(g) \quad \text{(reaction 2)}$$
$$2NO(g) + 2CO(g) \rightarrow N_2(g) + 2CO_2(g) \qquad \text{(reaction 3)}$$

(Here we've used C_7H_{16} as an example of an unburnt hydrocarbon, and NO as the main component of NO_x in car emissions.)

In these reactions, pollutants are being converted to CO_2, H_2O and N_2, which are all naturally present in the air. These reactions 'go' of their own accord, but under the conditions inside an exhaust system they go too slowly to get rid of the pollutants.

Assignment 11

Look at reactions 1–3.
a For each of the pollutants CO, C_7H_{16} and NO, say whether it is being oxidised or reduced.
b Which of these reactions would be important in controlling pollutants from
　i an ordinary engine?
　ii a lean-burn engine? (Look at Figure 28.)

A catalyst made of a precious metal such as platinum or rhodium speeds up these reactions in the exhaust system. Such catalysts are used in **catalytic converters**.

You can find out more about catalysts in **Chemical Ideas 10.5**.

A lean-burn engine uses an *oxidation catalyst system* which removes CO and C_xH_y. The exhaust gases are rich in oxygen, so CO and C_xH_y are oxidised to CO_2 and H_2O on the surface of the catalyst. This kind of catalyst system does little to remove NO because NO needs reducing, not oxidising, to turn it to harmless N_2. But that doesn't matter too much in a lean-burn engine because this type of engine produces less NO anyway (see Figure 28).

The three-way catalyst system
Catalytic converters can be fitted to ordinary engines too. But for an ordinary engine, a simple oxidation catalyst won't do – it wouldn't remove the NO, which is in much higher concentration in the exhaust than from a lean-burn engine. Here, a *three-way catalyst system* is needed which both oxidises CO and C_xH_y *and* reduces NO. This kind of catalyst system works *only* if the air–petrol mixture is carefully controlled so that it is exactly the stoichiometric mixture for the fuel. If the mixture is too rich then there is not enough oxygen in the exhaust fumes to remove CO and C_xH_y.

This means that cars fitted with three-way catalyst systems need to have oxygen sensors in the exhaust gases, linked back to electronically controlled fuel injection systems. You can see the arrangement in Figure 29.

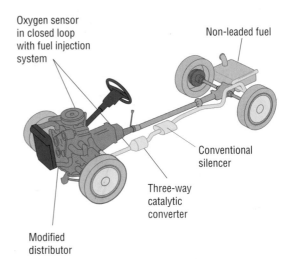

▲ **Figure 29** The three-way catalyst system.

Figure 30 shows the effect of catalysts on exhaust emissions from ordinary and lean-burn engines. A three-way catalyst system is the most efficient way of simultaneously lowering all three emissions.

All catalytic converters work only when they are hot. A platinum catalyst starts working around 240 °C, but you can get the catalyst to start working at about 150 °C by alloying the platinum with rhodium. These catalysts are poisoned by lead, so the converters can only be used with lead-free fuel.

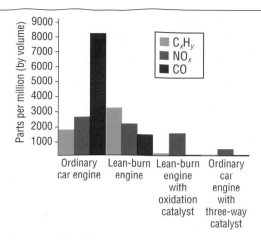

▲ **Figure 30** The effect of catalysts on the exhaust emissions from ordinary and lean-burn engines.

The catalyst is used in the form of a fine powder spread over a ceramic support whose surface has a network of tiny holes (see Figure 31). The surface area of the catalyst exposed to the exhaust gases is about the same as two or three football fields.

Assignment 12

a Why is it important that catalytic converters start working at as low a temperature as possible?

b What is meant by a *catalyst poison*?

c Why are these catalysts used in the form of a fine powder?

d Suggest a reason why a catalytic converter has to be replaced eventually.

e Catalytic converters convert the pollutants CO, C_xH_y and NO_x into harmless gases. This is still only a partial solution to the emissions problem. Why?

▲ **Figure 31** The honeycomb structure of this catalytic converter for a car exhaust gives it a high surface area.

DF8 *Using diesel fuel*

Table 6 Comparing pollutants from petrol and diesel cars.

Pollutant	Average in 2006 petrol car/g km^{-1}	Average in 2006 diesel car/g km^{-1}
CO_2	214	169
NO_x	0.034	0.317
particulates	0	0.028
CO	0.481	0.184

Table 6 shows that there is no clear cut environmental advantage in buying a modern diesel or petrol car.

For diesel engines, the most commonly used catalytic converter is the diesel oxidation catalyst. The catalyst uses the excess O_2 in the exhaust gas stream to oxidise CO to CO_2 and hydrocarbons to H_2O and CO_2. These converters are often very effective, virtually eliminating the smell of diesel and helping to reduce soot. However, in the presence of oxygen NO_x cannot be reduced and catalytic converters that reduce NO_x are under development.

Catalytic converters are unable to remove tiny carbon particulates. These can be partially removed by a diesel particulate filter (DPF).

Assignment 13

a Use the data in Table 6 to decide how *you* would choose between a petrol and a diesel car. (There is no 'right' answer to this question!)

b Write equations for the reactions in a diesel oxidation catalyst in which
 i   carbon monoxide and
 ii cetane ($C_{16}H_{34}$) are removed.

DF9 *Other fuels*

The other approach to tackling the emissions problem is to change the fuel used by cars.

Aromatic hydrocarbons make up as much as 40% of petrol. Aromatic hydrocarbons may cause higher CO, C_xH_y and NO_x emissions and some of them may cause cancer. Benzene is the worst and is strictly controlled, but others may also be controlled in the future.

Butane content too will probably be lowered in the future. Butane is volatile and is responsible for evaporative emissions leading to ozone formation and photochemical smogs.

However, both butane and aromatic hydrocarbons are high octane components of petrol. If they are removed then their octane quality must be replaced by something else. This is why the petrol companies looked to the oxygenates as a possible solution.

In the US, there is a very active policy to reduce the emissions caused by petrol itself and by its combustion products.

There are attempts to produce 'reformulated gasolines' that have:

- low volatility, thus reducing the concentration of hydrocarbons in the atmosphere;
- reduced concentrations of benzene (a carcinogen);
- added oxygenates (to improve burning and so reduce the concentration of pollutants such as CO).

All this has to be achieved without reducing the efficiency of the fuel.

Other hydrocarbon fuels

Liquified petroleum gas (LPG) comes from the distillation of crude oil and is often called 'autogas' when it is used in cars. It is a mixture of propane and butane in varying quantities, often around 60% propane. It has to be kept under pressure so that the hydrocarbons are stored as liquids. Petrol vehicles can be converted fairly easily to run on both fuels; one of the main changes needed is a larger fuel tank. Autogas has a high octane number and produces about 20% less carbon dioxide per mile than petrol. Because of the higher ratio of C : H it releases less CO. It also produces fewer unburnt hydrocarbons and NO_x than petrol. An advantage to vehicle owners is that road taxes and fuel taxes are lower than for petrol. The main disadvantage is that, although numbers are increasing, LPG filling stations are still relatively rare.

Liquid natural gas (LNG) is mainly methane and comes from oil and natural gas fields. Methane cannot

▲ **Figure 32** A bus powered by natural gas.

be liquefied by pressure alone and it must be cooled (below −160 °C) as well. This means that LNG is most suitable for large vehicles, especially as it works in modified diesel engines. Once again, there is a high C : H ratio so less CO is produced, and much less NO_x than diesel.

Biofuels

Ethanol is often added to petrol. It has a high octane number and is thought to be less polluting to the atmosphere. It is manufactured from ethene, which is obtained from the cracking of naphtha. However, ethanol can also be obtained from a renewable source – the fermentation of cane sugar juice. Brazil had to import oil during the 1970s but the climate there is good for growing cane sugar and a programme was set up to

▲ **Figure 33** A petrol station in Brazil. The fuel, gasohol, is a mixture of petrol and ethanol.

produce ethanol. By the 1990s Brazil could make enough ethanol to power about one-third of the country's 12 million cars – either using *gasohol* (a mixture of petrol and ethanol – Figure 33) or ethanol alone.

Burning ethanol produces less carbon monoxide, sulfur dioxide and nitrogen oxides than petrol, and it is believed that unburned ethanol does not contribute as much as hydrocarbons to photochemical smogs. The carbon dioxide produced is balanced by the absorption of carbon dioxide in growing sugar cane, and so using ethanol as a fuel adds less to the greenhouse effect.

Problems arise when oil prices drop because ethanol from fermentation is then less competitive. Large amounts of energy are needed for intensive cultivation of the crop, as well as large areas of land. There are queries about the overall energy efficiency of the process.

Biodiesel

Ordinary diesel engines can run on biodiesel. This is fuel made from vegetable oil or animal fat and almost any type can be used. For example, the waste oil used for frying chips can be used! Biodiesel is made through a process called transesterification. This converts vegetable oil and animal fat into esterified oil, which can itself be used as diesel fuel, or it can be mixed with regular diesel fuel. (You will learn about esters in **What's in a Medicine?**)

▲ **Figure 34** Biofuel research. Dr George Wake in his laboratory holding samples of a plant oil (dark brown, left) and biodiesel (yellow, right). The oil is extracted from Jatropha curcas seeds (bottom) and refined using equipment like that at right to produce the biodiesel.

$$\underset{\text{oil or fat}}{\begin{array}{c} H_2C-O-\overset{\overset{O}{\|}}{C}-R \\ | \\ HC-O-\overset{\overset{O}{\|}}{C}-R \\ | \\ H_2C-O-\overset{\overset{O}{\|}}{C}-R \end{array}} \quad + \quad \underset{\text{alcohol}}{3\ CH_3OH} \quad \xrightleftharpoons[\text{catalyst}]{\ OH\ } \quad \underset{\text{biodiesel}}{3\ CH_3O-\overset{\overset{O}{\|}}{C}-R} \quad + \quad \underset{\text{glycerol}}{\begin{array}{c} CH_2OH \\ | \\ CHOH \\ | \\ CH_2OH \end{array}}$$

Biodiesel is biodegradable so it is much less harmful than hydrocarbon oils to the environment if spilled.

Biodiesel has been shown to produce lower emissions than regular diesel fuel. The best thing about biodiesel is that it is made from plants and animals, which are renewable resources. It is sometimes described as 'carbon neutral' because the plants have absorbed the same amount of CO_2 in growing as is produced when the fuel is burnt. However, this ignores the energy used in the production process, which might have come from burning fossil fuels. Manufacturers also claim about a one-third reduction in carbon monoxide and hydrocarbon production and two-thirds reduction in particulates, compared with conventional diesel. NO_x emissions are sometimes quoted as being higher, though the almost total absence of sulfur compounds makes it easier to use catalytic converters to control this.

Assignment 14

a Write equations for the complete combustion of 1 mole of octane, cetane and ethanol.

b Look up the enthalpy changes of combustion of these compounds in Table 1 and Assignment 2 (page 24).

c Use this information and your equations from part **a** to work out the volume of CO_2 produced per kJ of energy transferred by the burning of each of the three fuels. (Assume 1 mole of CO_2 occupies 24 dm³.)

d Comment on the values you obtain.

DF10 *Hydrogen – a fuel for the future?*

Petrol has been developed to work well as a transport fuel. But problems remain – you have already studied the problem of emissions. Changing the fuel used by cars could help to solve the emissions problem. Figure 35 shows past and projected rates of world consumption of oil and coal. The use of petrol will decline in the future as oil supplies run out, so a search for alternative fuels will be vital to our transport system.

People who favour using hydrogen as a fuel see water as a plentiful source of hydrogen. If hydrogen could be extracted from water without consuming

fossil fuels it would reduce our dependence on these fuels, and help to reduce the amount of carbon dioxide released into the atmosphere.

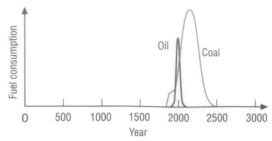

▲ **Figure 35** Past and projected rates of consumption of oil and coal. The use of fossil fuels will decline in the future as supplies run out.

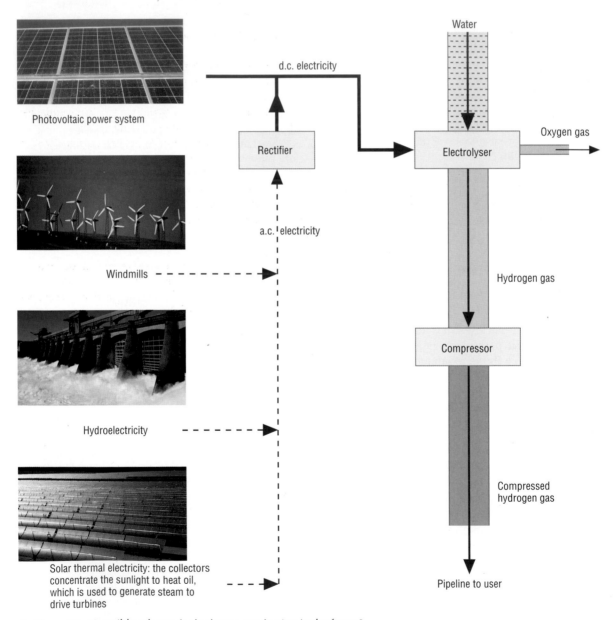

▲ **Figure 36** A possible scheme for hydrogen production in the future?

Hydrogen could be distributed as we now distribute natural gas, and burned as a heating fuel, used in internal combustion engines or converted into electricity in fuel cells.

But how can we make the hydrogen? The most likely large-scale method of producing it seems to be by electrolysis of water, obtaining the energy needed for this from some renewable source such as solar cells. But why produce electricity and then use it to make hydrogen? There are two major advantages:

- hydrogen can be stored;
- it can be used in the internal combustion engine.

How can we generate the electricity needed to make hydrogen by electrolysis? The present methods of generating electricity are:

- Burning the fossil fuels coal, oil and natural gas – oil and natural gas, of course, could be used to make fuels for vehicles. Also, all of these fuels produce carbon dioxide, as well as NO_x and SO_2, so there is hardly any point in making hydrogen this way. Burning biofuels is worthwhile to generate electricity for other uses but is not worth it for making hydrogen.
- Nuclear power – this does not produce carbon dioxide, NO_x or SO_2 but there are risks of radioactive isotopes being released. Also, the spent fuel rods are highly radioactive and take hundreds of years to become safe so there are major problems in 'decommissioning' a nuclear power station based on fission reactors. Once *fusion* power stations come on line then most of these problems will be solved, but this is still many years away.
- 'Alternative' energy – this includes photovoltaic cells, windmills, hydroelectricity, solar thermal energy and wave barrages. These must be exploited as much as possible, but many of them are unsightly and do not currently produce a large, reliable supply of electricity.

A possible scheme is shown in Figure 36. All the technologies needed already exist. We have large-scale electrolysis plants for other purposes, large storage tanks and a network of gas pipelines.

▲ **Figure 37** A car powered by a hydrogen fuel cell.

The **hydrogen economy** would use hydrogen as a way of storing and distributing energy. If systems are costed over whole lifetime use in terms of money and energy, then distributing hydrogen by pipeline may be cheaper than transmitting electricity.

Fuel cells are being used to generate electricity on a small scale in cars (see Figure 37).

The main problem in the design of such a car is the storage of hydrogen. A large volume of gaseous hydrogen is required to get the mileage equivalent to a fuel tank of petrol. We need to find some way of storing it more compactly – one solution is to store it as a liquid in a high-pressure fuel tank.

Schemes for developing alternative fuel use on a large scale depend on long-term and large-scale investments in new infrastructures – and so success will depend on political as well as economic factors. The reward could be cleaner renewable fuels.

You can make decisions on fuels for the future by doing **Activity DF10**.

Assignment 15

The petrol tank of a typical car holds about 45 litres of petrol (approximately 10 gallons).

a Calculate the amount of energy released by burning 45 litres of petrol. Use the following information to help you.
  - Assume that petrol is octane (C_8H_{18}).
  - The standard enthalpy change of combustion of C_8H_{18} is $-5500\,kJ\,mol^{-1}$.
  - The density of octane is $0.70\,g\,cm^{-3}$.
  - 1 litre is $1000\,cm^3$.

b Calculate the mass and volume (at 20°C and 1 atmosphere pressure) of hydrogen needed to provide the same amount of energy as 45 litres of octane.
  - The standard enthalpy change of combustion of H_2 is $-286\,kJ\,mol^{-1}$.
  - 1 mole of a gas at 20°C and atmospheric pressure has a volume of about 24 litres.

You can remind yourself about calculations involving gases by reading **Chemical Ideas 1.4**.

Hydrogen engines are efficient. In motorway driving conditions a hydrogen engine can be over 20% more efficient than a petrol engine. In city 'stop-go' driving conditions the hydrogen engine is about 50% more efficient than a petrol engine.

c Taking efficiencies into account, what mass and volume of hydrogen are needed to give the same mileage as 45 litres of petrol in
  i motorway driving conditions?
  ii city driving conditions?

DF11 *Summary*

This module began by considering some of the desirable properties of a good fuel. This led you to compare the energy released when different fuels are burned in oxygen and to a general study of energy changes in chemical reactions. You then found out how petrol and diesel are made and how chemists and chemical engineers are helping to develop better fuels for motor vehicles. This means producing fuels which give improved performance and greater fuel economy, so that we use our precious reserves of crude oil as economically as possible.

Petrol and diesel are mixtures of hydrocarbons, many of which are alkanes. This led you to find out more about the properties and reactions of alkanes. But the hydrocarbons that give the best performance in a petrol engine are not the ones that are most plentiful in crude oil. So, after the primary distillation of crude oil into fractions, a whole range of chemical processes are carried out in the refinery to 'doctor' the hydrocarbons to suit our needs.

Petrol must have the correct octane number to avoid auto-ignition. Branched-chain alkanes have higher octane numbers than straight-chain alkanes, which led you to study structural isomerism in more detail. For some molecules it is enough to rearrange the positions of the atoms, but large molecules must be broken down into more useful smaller ones, and some smaller ones are joined together. The final petrol is a blend of many hydrocarbons, and possibly some oxygenates too. This led you to find out about the structures of compounds in other homologous series: alkenes, arenes, alcohols and ethers. The mixing of fuel components to form the final 'blend' introduced you to simple ideas about entropy.

You went on to consider the problems of emissions from motor vehicles, the difference between primary and secondary pollutants and the formation of photochemical smog. You examined ways in which car manufacturers and oil companies are tackling the problem of emissions – by changing engine technology to control the way the fuel burns, by changing the composition of the fuel and by using catalytic converters to speed up reactions in the exhaust that convert the pollutant to less harmful gases. You then looked in more detail at the role of catalysts in chemical reactions.

However, the story does not end with petrol and diesel. Our supplies of crude oil are finite and are needed for more than fuels alone. You then explored a range of alternative fuels, such as hydrogen, for motor vehicles for the future.

Much of the chemistry you have covered in this module is fundamental to other areas and you will need to use these ideas in other parts of the course.

Activity DF11 will help you to check that you understand the ideas in this module.

ELEMENTS FROM THE SEA

Why a module on 'Elements from the Sea'?

The first module in the course – **Elements of Life** – told the story of how the elements were formed. The theme is taken further in this module, which tells how we have learned to win back some elements from the natural resources that contain them and turn them into useful substances. The halogens and their compounds, some of which are present in sea water, have been chosen.

The module introduces two major types of inorganic chemical reaction – redox and ionic precipitation. It also introduces more organic chemistry by considering the formation and reactions of halogeno-alkanes. Some important inorganic halogen and redox chemistry is also covered.

Areas of chemistry that you met in earlier modules are revisited and taken further. The concept of amount of substance is extended to include concentrations of solutions. Ideas about the electronic structure of atoms are developed to include the distribution of electrons in atomic orbitals, and the properties of covalent compounds are linked to their bonding and structure, including some concepts of intermolecular bonding.

Overview of chemical principles

In this module you will learn more about ideas you will probably have come across in your earlier studies:
- the halogens
- ions in solids and solutions
- dissolving
- acids
- precipitation.

You will learn more about some ideas introduced in earlier modules:
- amounts in moles (**Elements of Life**)
- electronic structure of atoms (**Elements of Life**)
- covalent bonding (**Elements of Life** and **Developing Fuels**)
- the Periodic Table (**Elements of Life**)
- ionisation enthalpies (**Elements of Life**).

You will also learn new ideas about:
- redox reactions
- intermolecular bonds
- methods of manufacturing chemicals industrially
- halogenoalkanes.

ELEMENTS FROM THE SEA

ES1 *Why is the sea so salty?*
A story of smokers and solutions
Black smokers

In Yellowstone Park one of the main attractions is a geyser called Old Faithful. This geyser spouts a column of steam and hot water into the air several times a day. Hydrothermal vents are geysers that occur on the sea floor. These vents constantly spew superhot mineral-rich water that helps to support a diverse community of organisms.

The Earth's crust, including the sea floor, is made from large pieces of solid rock called plates. These rest on the partially molten mantle, which contains radioactive elements. As these elements decay, a large amount of energy is released, heating the surrounding rocks. The hot rock rises, creating a convection current that makes the plates of the crust move. Hydrothermal vents form where the plates of the Earth's crust move apart, producing cracks in the sea floor. Sea water enters the cracks and is heated by the molten rock of the mantle. The hot water dissolves chemicals in the rocks producing a concentrated solution of minerals. The hot solution then rises and goes back into the sea, taking the dissolved minerals with it. Figure 1 shows a plate margin rift in Thingvellir, Iceland – this plate boundary is a part of the mid-Atlantic ridge. The plates on either side of the boundary are moving apart, allowing new land and ocean floor to form.

Of the hydrothermal vents produced in this way 'black smokers' are the hottest. The mixtures they give out contain a large amount of iron and compounds of sulfur – the presence of iron sulfide gives the smoker its black colour.

These insoluble materials are often formed by a process known as **precipitation**. In solution are compounds of chlorine, bromine (and a variety of metals, including some from Groups 1 and 2).

You have already seen how to write balanced equations. During precipitation reactions insoluble materials are formed. You can learn about this and how to write ionic equations to represent these reactions in **Chemical Ideas 5.1**.

▲ **Figure 1** A plate margin rift.

▲ **Figure 2** A black smoker vent.

Hydrothermal vents have now been discovered in an unexpected part of the ocean floor, as described in an article written in *Science News* by scientists at the University of Washington (see box on page 50).

Hydrothermal vents account for much of the dissolved chemicals present in sea water. Other dissolved chemicals have another source – as rainwater passes through soil and percolates through rocks, it dissolves some of the minerals; a process called weathering. Eventually this water, with its small amounts of dissolved minerals or salts, reaches streams and flows into lakes and the ocean. Over time, and with

NEW TYPE OF HYDROTHERMAL VENT LOOMS LARGE

A hydrothermal vent system, dubbed the Lost City, rises on an undersea mountain in the Atlantic Ocean. The mineral deposits produced by these vents include, at 18 stories tall, the largest ever observed. The deposits rise like chimneys above the sea floor and are host to rich communities of microbial life. The hydrothermal field was discovered near the top of a seamount called the Atlantis massif. This submerged mountain lies about 2500 km east of Bermuda.

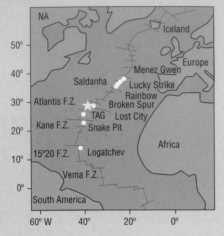

▲ **Figure 3** The location of the Lost City field (shown as a yellow star) and the ridge-top hydrothermal vent fields (shown as dots).

▲ **Figure 4** A photomosaic of a carbonate chimney typical of those found at the Lost City hydrothermal field.

Hydrothermal activity at the Lost City vents appears to be fuelled by a chemical reaction taking place in the seamount's rocks. Those rocks, which were once deep within the Earth's mantle, were exposed by erosion after the seamount rose to near or above the sea's surface. Now in contact with sea water that circulates through fissures in the rock, a mineral called olivine changes into another mineral called serpentine. This reaction generates heat and produces a number of substances including methane and hydrogen – these gases were identified in the highly alkaline water that spouts from the Lost City vents at temperatures between 40 °C and 75 °C.

When the heated water emerges from the vents into the cool ocean water, chemical reactions deposit vast amounts of minerals and form an unearthly landscape of chimneys, mounds and spires. Analysis of the deposits shows a mixture that includes calcium carbonate – the same white mineral that forms stalactites and stalagmites in terrestrial limestone caves.

Because calcium carbonate is much stronger than the sulfide minerals typically formed by other hydrothermal vents, the Lost City deposits can grow exceptionally high. The tallest of the 30 or so carbonate structures, a 60 m chimney, is still growing. It easily tops the largest chimney ever found at a black smoker – a 41 m tall sulfide specimen off Washington State that broke in half and toppled over.

The concentrations of chemicals in the warm, alkaline water gushing from the Lost City vents are similar to those that some scientists contend may have been present in similar hydrothermal systems during the early years of Earth's history. Hydrothermal activity plays an important part in moving dissolved minerals and other substances through the Earth's crust and into the ocean. There is enough water flowing through such sea floor vent systems to recycle the ocean's volume every 500 000 to 1 million years.

The methane, hydrogen and minerals emitted by the Lost City vents support a thriving community of microbes. Although a few crabs and sea urchins walk the streets of the Lost City, these complex deep-sea creatures don't seem to be closely associated with the vent system. The Lost City appears to be devoid of other multicellular animals, such as the tube worms that populate some black smoker sites.

Science News (July 2001)

the constant evaporation of water from the surface of the oceans, this has led to an accumulation of dissolved salts in the oceans of the world. Most seas around the world have similar concentrations of dissolved minerals, but there are some exceptions, such as the Dead Sea.

Different minerals dissolve in different concentrations in the oceans. You can find out how to calculate concentrations of solutions in **Chemical Ideas 1.5**.

ES2 *The lowest point on Earth*

The Dead Sea is the lowest point on Earth, almost 400 m below sea level in the Rift Valley, which runs from East Africa to Syria. It is like a vast evaporating basin – water flows in at the north end from the River Jordan, but there is no outflow. The countryside around it is desert and in the scorching heat so much water evaporates that the air is thick with haze, making it hard to see across to the mountains a few kilometres away on the other side. This steady evaporation of water for thousands of years has resulted in huge accumulations of salts, so that the water in the Dead Sea is much denser than elsewhere.

The Dead Sea is such a natural curiosity that surveys of the salt concentration were conducted as early as the seventeenth century, even though many of the elements in the salts were then unknown.

The water contains about 350 g dm^{-3} of salts compared with 40 g dm^{-3} in water from the oceans. A comparison of the most abundant ions present in Dead Sea water and typical ocean water is shown in Table 1.

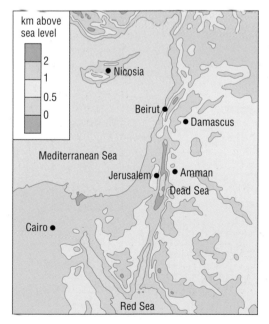

▲ **Figure 5** The region around the Dead Sea in relief.

▲ **Figure 6** The Dead Sea – humans float easily due to the high density of the water.

Table 1 Compositions of samples of typical ocean water and Dead Sea water.

Ocean water		Dead Sea water	
Ion	Mass in 1 dm³ of water/g	Ion	Mass in 1 dm³ of water/g
Na⁺	11	Na⁺	39
K⁺	0.4	K⁺	6.9
Mg²⁺	1.3	Mg²⁺	39
Ca²⁺	0.4	Ca²⁺	17
Cl⁻	19	Cl⁻	208
Br⁻	0.07	Br⁻	5.2
HCO₃⁻	0.1	HCO₃⁻	trace
SO₄²⁻	2.5	SO₄²⁻	0.6

Estimates suggest that there are about 43 billion tonnes of salts in the Dead Sea, and a particular feature is the relatively high proportion of bromides.

The sea is the major source of minerals in the region. A chemical industry has grown up around the Dead Sea in Israel and it has become one of the largest exporters of bromine compounds in the world. The annual production of bromine compounds in Israel exceeds 230 000 tonnes.

The solubility of an ionic compound depends, in part, on the charge on its ions. You can see from Table 1 that different ions carry different charges. You can find out about ionisation of atoms of an element to form ions in **Chemical Ideas 2.5**.

You can learn more about ionic compounds as solids and what happens to ionic compounds when they dissolve to form solutions in **Chemical Ideas 5.1**.

You will learn how to work out the formula of an ionic compound in **Activity ES2.1**, and **Activity ES2.2** investigates some of the ideas about dissolving and precipitation that you will be reading about in this module.

You can find out about writing formulae in **Chemical Ideas 3.1**.

Assignment 1

Dead Sea water is certainly more salty than ocean water, but Table 1 shows that there are differences in ionic composition between the two.

a   What do you notice about the abundances of the ions of Group 1 and Group 2 elements in the two samples?

b   Compared with ocean water, Dead Sea water has a particularly high proportion of one ion. Which ion is this? How many times more abundant is this ion in Dead Sea water than in ocean water?

c   The ions in the Dead Sea water come from ionic compounds that have dissolved in the water. Work out the formula of these ionic compounds, using the information in the table:
  i   sodium sulfate
  ii calcium hydrogencarbonate
  iii magnesium bromide.

Bromine from sea water

The levels of bromide ions in the Dead Sea are high compared to many other natural water sources. This makes it an ideal source of bromine. Bromine is a Group 7 element. In the past it has been used in the manufacture of 1,2-dibromoethane – used in leaded petrol to remove lead deposits from engine cylinders during combustion. Silver bromide was used in photographic film as it is light sensitive, in common with other silver halides. Nowadays bromine and organic bromo-compounds have a wide range of uses in the pharmaceutical industry (e.g. in the manufacture of analgesics, sedatives and antihistamines). Currently pharmaceuticals that use bromine in their manufacture are undergoing trials for treatment of Alzheimers disease. Bromo-compounds are used in the manufacture of brominated flame-retardants (BFRs), which have been instrumental in saving many thousands of lives.

Turning dissolved bromide ions into bromine involves simple chemistry – it is easy to do in the laboratory just by adding chlorine to a solution containing bromide ions. Industrially it is more complicated, and involves the use of some ingenious engineering as you can see in Figure 7.

The Dead Sea Bromine Group Ltd at S'Dom in Israel opened on the southwest shore of the Dead Sea in the 1930s and is still producing bromine.

Partially evaporated acidified sea water is warmed, and then chlorine is added to displace bromine from the bromide ions. Sea water is slightly alkaline and it must be acidified with sulfuric acid to lower the pH before the chlorine is added – both chlorine and bromine react with water at higher pH values. Bromine is volatile (boiling point 331 K) and bromine vapour is given off, with water vapour, when steam is blown through the solution. The vapours are then condensed and two layers form because liquid bromine is not very soluble in water. The dense bromine layer is run off from the water that floats on the top. The impure bromine is then distilled and dried.

The chlorine needed in the process is produced by electrolysis on site. You will look at the electrolysis of brine to produce chlorine in **Section ES3**.

The reaction of chlorine with bromide ions is an example of a very important type of chemical reaction in which both oxidation and reduction take place – such processes are called **redox reactions**. Many reactions of the halogens are redox reactions.

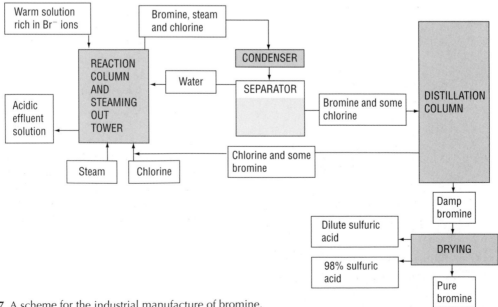

▲ **Figure 7** A scheme for the industrial manufacture of bromine.

Assignment 2

a In the process described above, bromine is separated from other materials involved in the process. Which properties of bromine make it possible to separate it from
   i   water
   ii chlorine?

b Write an equation for the reaction of chlorine with bromide ions to produce chloride ions and bromine.
c In the production of 1 tonne of bromine, what mass of chlorine is required?
d What volume would this mass of chlorine gas occupy at standard temperature and pressure?

Chemical Ideas 9.1 tells you about redox reactions.

If you need help with Assignment 2 you might like to refer back to **Chemical Ideas 1.2**, **1.3** and **1.4**.

The production of bromine depends on the different reactivities of the halogens – you can learn more about these trends in **Chemical Ideas 11.4**.

Chemical Ideas 2.4 tells you more about the energy levels in atoms, and how electrons are arranged in these energy levels. This will explain why the halogens are in the p-block of the Periodic Table and why elements are able to form compounds in which they have a certain oxidation state.

You can revise the classification of the Periodic Table, including p-block elements, in **Chemical Ideas 11.1**.

Assignment 3

Many elements, like chlorine, exhibit a range of different oxidation states. One way of summarising these is to construct an oxidation state chart – this shows the oxidation states assigned to the element and some of its compounds. An example of this is shown in Figure 8 for arsenic.

▶ **Figure 8** Oxidation state chart for arsenic.

a Work out the oxidation state of the chlorine in each of these species.
   ClO_2 ClO_3 $HClO_4$
   $MgCl_2$ Cl_2O_7 HCl
   $HClO_2$ Cl_2O Cl_2
b Use your answers to draw an oxidation state chart for chlorine.

ES3 *An industrial case study – how best to manufacture chlorine?*

The human demand for chlorine is high and it is manufactured worldwide on an enormous scale. Chlorine can be made by the electrolysis (decomposing a compound using an electric current) of a concentrated solution of sodium chloride (sometimes called brine).

Although the sea contains high concentrations of sodium chloride, almost pure sodium chloride is found as rock salt. It can be recovered either by underground mining or by pumping water into the salt and collecting the salt solution at the surface. The electrolysis of brine also generates hydrogen and sodium hydroxide (an alkali). As chlorine manufacture and sodium hydroxide manufacture are directly linked, we often talk about the chlor–alkali industry when referring to the production of these chemicals.

The co-products, sodium hydroxide and hydrogen, can be sold and this helps to reduce the waste, as well as increasing profitability.

You can learn more about the factors that chemical companies need to take into account when setting up and running a chemical plant in **Chemical Ideas Sections 15.1– 15.6**.

While reading the following section of this module, you should consider how each of the processes is influenced by cost, raw materials, safety, siting of the plant and waste disposal.

There are three common technologies for producing chlorine from brine – the mercury cell, the diaphragm cell and the membrane cell. We will consider two of them in detail – the mercury cell and the membrane cell. Both produce chlorine by electrolysis.

▲ **Figure 9** A bank of mercury cells in a chlor–alkali plant.

The mercury cell

The mercury cell was the first large-scale method used for making chlorine, but it is gradually being phased out. This is because the mercury cell is responsible for losses of toxic mercury to the environment, and because it is more expensive to run than other types of cell. Although there has been a large decrease in the amount of mercury given out to the environment, mercury emissions from chlorine manufacture still accounted for an average of 1.05 g per tonne of chlorine from all the mercury cell plants in Western Europe in 2006. Concerns over the effects of mercury on the environment have driven European chlorine manufacturers to introduce voluntary phasing out of the remaining mercury-based chlorine plants by 2020. Similar conversions to membrane or diaphragm cells are taking place around the world, either on a voluntary basis or because of changes in legislation.

The membrane cell

The membrane cell is the most modern of the electrolysis methods. It was introduced in the 1980s and many companies producing chlorine have replaced their old mercury cells with membrane cells.

Although there are high initial start-up costs for building new cells, the membrane cell has the following advantages over the older mercury-based technology:

▲ **Figure 10** A bank of membrane cells for chlorine production.

- the running costs for the membrane cell are lower, largely due to lower energy requirements per tonne of chlorine produced;
- a cell with a much larger capacity for chlorine production can be installed in the space that held a mercury cell, allowing a greater quantity of chlorine to be made at the plant;
- it is not necessary to carry out further costly processing to remove mercury from the products;
- there is less environmental pollution.

These differences mean that the start-up costs for installing a membrane cell (which form part of the fixed costs for the operation) can be recouped within about five years of its introduction.

The half-equations involved in the electrolysis of sodium chloride solution are:

- at the positive electrode $2Cl^-(aq) \rightarrow Cl_2(g) + 2e^-$
- at the negative electrode
$$2H_2O(l) + 2e^- \rightarrow 2OH^-(aq) + H_2(g)$$

The equation representing the overall reaction occurring in the cell is

$$2Cl^-(aq) + 2H_2O(l) \rightarrow Cl_2(g) + H_2(g) + 2OH^-(aq)$$

Assignment 4

a Calculate the amount (in moles) of sodium hydroxide, NaOH, in 1 tonne of solid sodium hydroxide.
b What amount (in moles) of chlorine, Cl_2, is produced for each mole of NaOH?
c Calculate the mass of chlorine produced at the same time as 1 tonne of sodium hydroxide.

The electrolysis cell used to make chlorine, hydrogen and sodium hydroxide by electrolysis of sodium chloride solution must be designed to:

- prevent the chlorine produced at the positive electrode from reacting with the hydroxide ions around the negative electrode;
- minimise chloride ions diffusing into the solution around the negative electrode, which would contaminate the sodium hydroxide solution;
- minimise hydroxide ions being lost by diffusion away from the negative electrode towards the positive electrode;
- prevent mixing of chlorine and hydrogen, which could lead to an explosion.

These requirements are achieved by using a membrane that acts as a barrier to all gas and liquid flows and allows only the transport of charged sodium ions between compartments. Sodium ions in hydrated form ($Na^+(aq)$ pass through) so some water is transferred, but the membrane is impermeable to free water molecules.

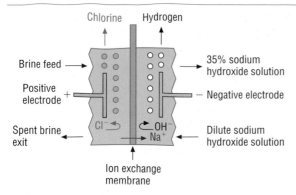

▲ **Figure 11** The changes occurring in a membrane cell for the manufacture of chlorine, hydrogen and sodium hydroxide.

The membranes are based on poly(tetrafluoroethene), usually referred to as PTFE or Teflon, which is resistant to high temperatures and to chemical attack. The PTFE is modified so that it contains negatively charged side chains. These attract the positively charged sodium ions, which pass through the membrane, attracted by the charge on the negative electrode. Negative ions, such as OH^-, are repelled by the side chains and this stops these ions from passing through the membrane.

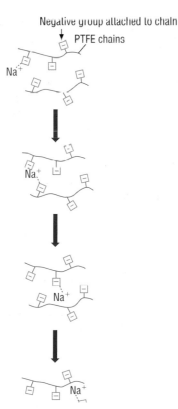

▲ **Figure 12** Sodium ions passing through a channel in an electrolysis cell membrane.

ES4 *From atomic bombs to safer drinking water*

There are a variety of risks associated with the production and transportation of halogens, yet they make a valuable contribution to improving our lives. We need to weigh up the risks and benefits associated with each halogen and its associated compounds before we decide to use them.

From safer water to cleaner clothes

As well as its use in extracting bromine from sea water, chlorine is used to make bleaches. Chlorine gas is passed through a cold solution of sodium hydroxide. The sodium hydroxide solution reacts with the chlorine to form sodium chlorate(I).

$$2NaOH + Cl_2 \rightarrow NaCl + NaClO + H_2O$$

As both reactants are produced from the electrolysis of brine, a plant producing chlorine could also have a bleach production facility on the same site.

A solution of about 12% NaClO by mass is used in some water purifying plants to kill bacteria. A solution of about 5% is used in household bleach products, often to remove stains from clothing. The bleach, which is an oxidising agent, removes stains by breaking bonds in coloured chemicals to form colourless products.

In **Activity ES4.1** you will use a titration to compare a range of bleaches.

Chlorine has a poor public image. It is associated with pollution – pollution of the land through pesticides that contain organochlorine compounds and pollution of the upper atmosphere through CFCs (chlorofluorocarbons). You may also have heard about the use of chlorine as a poisonous gas. Both chlorine and a compound derived from it, phosgene, $COCl_2$, were used with deadly effect in the trenches in the First World War. They are also thought to have been used against civilian populations in recent years.

However, chlorine is used in many ways to make our lives safer and more comfortable. About 50 million tonnes of chlorine are produced worldwide annually. The best known use is in water treatment, where it is added to the water to kill bacteria and other pathogens.

▲ **Figure 13** A technician dropping a 'chlorine tablet' into a swimming pool dispenser.

Assignment 6

a   Calculate the oxidation state of each of the elements in all of the chemicals before and after the reaction represented by this equation:

$$2NaOH + Cl_2 \rightarrow NaCl + NaClO + H_2O$$

b   This is an example of a redox reaction. Why is it an unusual one?

A large quantity of chlorine is used to make organic chemicals such as chloroethene, which is used to make poly(chloroethene) or PVC. It is needed for the manufacture of polyurethanes and is present in a wide variety of solvents (such as trichloroethene) used both in the dry cleaning of clothes and in industry to clean grease off metals. The production of organic compounds of the halogens will be considered in more detail in **Section ES5**. You can find out more about the formation of polymers like PVC in **Polymer Revolution**.

▲ **Figure 14** A child crash test dummy made from polyurethane.

Uses of the other halogens

Use of the halogens, in their elemental form, is linked with their physical states. Fluorine and chlorine are gases at room temperature and pressure. Bromine is one of the few elements that is in the liquid state under these conditions, while iodine and astatine are both solids.

Chemical Ideas 3.1 contains a useful section on electronegativity and bond polarity.

The physical state of the halogens and other chemicals at room temperature is explained in terms of intermolecular bonds in **Chemical Ideas 5.3**.

You will find out more about the physical states of the halogens in **Activity ES4.2**.

Like chlorine, the public considers most of the halogens to be dangerous chemicals because they are all toxic in their elemental forms. However, they all play an extremely important role in our everyday lives, being used to make a variety of beneficial products.

The most reactive halogen

Fluorine was the last of the halogens to be isolated because it is the most reactive member of the group, and so the most difficult to isolate. The first large-scale production of fluorine was for the atomic bomb project, known as the Manhattan project, during the Second World War – fluorine was needed to make uranium hexafluoride.

Because it is the most reactive of the halogens, fluorine is extremely dangerous to use. It is so reactive that it is almost impossible to store because it reacts with the chemical used to make the container. Instead it is necessary to generate the fluorine and then use it immediately to make the required product.

Fluorine is made from the mineral fluorite, which contains calcium fluoride. A crystal of calcium fluoride is shown in Figure 15. The calcium fluoride is reacted with sulfuric acid, producing hydrogen fluoride.

$$CaF_2 + H_2SO_4 \rightarrow CaSO_4 + 2HF$$

▲ **Figure 15** Fluorite crystals

The hydrogen fluoride is liquefied and electrolysed to produce fluorine and hydrogen. Some fluorine is reacted with sodium to form sodium fluoride – this is added to toothpaste and, in some cases, domestic water supplies, to help to strengthen tooth enamel and prevent dental decay. Some people object to the addition of fluoride ions to drinking water because this takes away their choice to drink fluoridated water or not. What do you think?

Fluorine is also used to make a wide range of organic compounds. These include hydrochlorofluorocarbons (HCFCs) used for air conditioning and refrigeration, and the polymer PTFE, which can be used to make non-stick cookware – it is also used in reconstructive and facial surgery. You can find out more about HCFCs in **The Atmosphere** and PTFE in **Polymer Revolution**.

Dark, dense and dangerous

Bromine was discovered in 1826 by a French chemist, Jérôme Balard, who later became a professor at the Sorbonne in Paris. His discovery helped Johann Döbereiner to spot the idea of 'triads', which was a step on the way to the development of the Periodic Table.

Bromine has a dense and choking vapour – hence its name, which is based on a Greek word, 'bromos', meaning stench. The liquid produces painful sores if spilled on the skin – you will use bromine several times in this course and you must take great care when you handle it.

Great care has to be taken when transporting bromine (Figure 16). Most of it is carried in lead-lined steel tanks supported in strong metal frames – each tank holds several tonnes of the element. International regulations control the design and construction of road and rail tankers.

▲ **Figure 16** A road tanker for the transport of bromine. Note the 'cage' to give extra protection in the case of an accident.

The industry's safety record is good, but there have been some accidents. One such accident is discussed in Figure 19 and Assignment 7 (page 58).

Activity ES4.3 looks in detail at how bromine is handled when it arrives at a chemical plant.

Bromine may once have been just a laboratory curiosity, but it is now an important industrial commodity manufactured on a large scale (**Section ES2**). One of its most important uses is in the production of flame retardants (Figure 17, page 58). For example, tetrabromobisphenol A (TBBA) can be incorporated into some polymers (via the –OH group) making the polymer much less prone to combustion. Silver bromide was very important in traditional photography, although this use is in decline due to an increase in the use of digital photography. Bromine is also used as a reactant in the synthesis of medicines, dyes and pesticides.

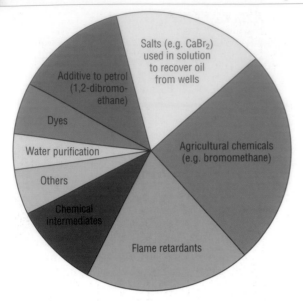

▲ **Figure 17** A pie chart showing the uses of bromine.

Bromine compounds are used in agriculture. Bromomethane (CH_3Br) is used as a fumigant against many pests (Figure 18).

Unfortunately, if bromomethane gets into the upper atmosphere it forms bromine atoms which destroy ozone. (You can find out more about ozone depletion in **The Atmosphere**.) For this reason, attempts are being made to phase out bromomethane and other fumigants.

▲ **Figure 18** Bromomethane is a very effective fumigant to protect food, grain and wood from pests, such as the Asian long-horned beetle, which can devastate them in storage.

Assignment 7

According to *The Jerusalem Post* report (Figure 19), a safety expert recommended more careful monitoring of the transportation of dangerous chemicals. In the light of the report, what do you think the chemical industry does to reduce the risk of injury or death when chemicals such as bromine are transported?

CHEMICAL SCARE SHAKES NEGEV

The driver of the truck that overturned near here early yesterday morning, sending an orange cloud of poisonous bromine gas into the sky over the Arava, said before he died that the accident occurred because he fell asleep at the wheel. The cloud of gas formed by the evaporating bromine, which had a diameter of some 10 km later in the morning, forced scores of residents to leave their homes for several hours. Negev police closed off the road for the day.

According to a senior member of the Dead Sea Works rescue crew, the driver, Yisrael Taib, 32, of Dimona, was trapped when his flatbed truck flipped over on its side at about 5.15 am after rounding a bend some 20 km south of the Arava junction.

Eight people were injured as a result of the accident, most of them rescue workers who tried frantically but in vain to extricate Taib, who was crushed by the steering wheel and the engine block in the mangled cabin.

A senior police officer on the scene praised the rescue crew for the speed with which it reached the spot.

The rescuers arrived shortly after 6 am and worked for several hours, even after a wind change sent the poisonous gas straight towards them.

A safety expert from a major industrial concern in the Negev told *The Jerusalem Post* that the bromine had apparently been packed according to international safety standards, but that yesterday's accident highlighted the need for more careful monitoring of the transportation of dangerous chemicals.

The expert, who sits on a national safety committee, told *The Post* that repeated efforts over the years to have 'black boxes' installed in the cabins of trucks carrying dangerous materials had been thwarted by trucking officials. The expert, who asked not to be identified, conceded that the black boxes wouldn't have prevented yesterday's accident, but would have told safety officials how fast the driver was travelling and how many hours he had been working.

The truck, on its way from the Dead Sea bromine plant south of Beersheva to the port in Eilat, was carrying two containers with over 20 tons of liquid bromine.

They said it was the first accident ever in Israel involving bromine, a deep rust-coloured liquid used in industry, agriculture and medicine. The bromine evaporated quickly in the desert heat. Bromine's boiling point is 58 °C.

▲ **Figure 19** Report from *The Jerusalem Post*.

The tangle o' the isles

Just below the surface of the seas off the coast of Scotland lie hidden forests of the seaweed called kelp. This has historically been a valuable source of organic matter that local people used to fertilise their soil. In the early part of the nineteenth century, Napoleon Bonaparte was looking for a source of nitrate for making explosives. In 1811, the French entrepreneur Bernard Courtois responded to Napoleon's demands by attempting to make nitrate from rotting seaweed. He noticed some curious purple fumes and, by chance, discovered another of the halogens – iodine. This led to a whole new industry developing in the coastal areas of Scotland.

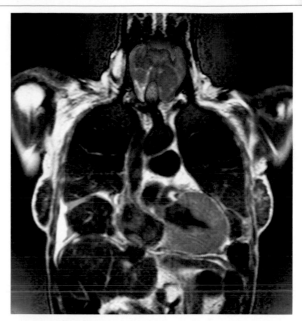

▲ **Figure 21** This MRI scan shows in pink the swollen thyroid gland typical of someone suffering from goitre.

▲ **Figure 20** Kelp washed up on the coast in Northumberland.

Iodine is a dark grey solid at room temperature, which forms a violet coloured vapour when it is heated (hence its name, from the Greek word, 'iodes', which means violet). Apart from seaweed, iodine can be obtained from some sources of brine, in a similar way to that described earlier for the production of bromine by treating the brine with chlorine.

Iodine is used as an antiseptic solution, which is produced by dissolving iodine in ethanol, and for making medicines. Humans need iodine in order to remain healthy as it is used in our bodies to produce hormones in the thyroid gland. A lack of iodine can lead to swelling of the thyroid gland, and a condition known as goitre results (Figure 21). We get iodine from our food – often from vegetables that have been grown in soil containing iodide ions.

In **Chemical Ideas 11.4** you can find out about the chemistry of the halogens.

You can investigate some of the properties of bromine and other halogens and their compounds in **Activity ES4.4**. Also, **Activity ES4.5** will help you to check your notes on the first part of this module.

ES5 *Hydrochloric acid – an industrial success*

Hydrochloric acid can be made by a variety of methods. In many cases, it is a secondary product – a process is carried out to make a particular chemical, and the plant uses a co-product or by-product to generate hydrogen chloride for hydrochloric acid manufacture.

A simple method for the production of hydrochloric acid is to start by making hydrogen chloride gas directly from the elements. This would be possible at a plant producing chlorine from brine by electrolysis, as described in **Section ES3**, as both chlorine and hydrogen are produced in the process

$$H_2 + Cl_2 \rightarrow 2HCl$$

This is a good example of atom economy. For more information on this see **Chemical Ideas 15.7**.

Once the hydrogen chloride has been made, it is dissolved in water to produce hydrochloric acid at a concentration of about 30% by mass. It is more economical to transport it in the concentrated form and then dilute it, if necessary, at its destination.

You can learn about calculating the concentration of a solution in **Chemical Ideas 1.5**.

▲ **Figure 22** A plant for manufacturing hydrochloric acid.

You can find out about how a titration can be used to find the concentration of acid in a solution in **Activity ES5.1**.

A large proportion of the hydrochloric acid that is made is a co-product from the chlorination of organic compounds. For example, the first stage in the manufacture of poly(chloroethene) (or polyvinyl chloride, PVC) is the reaction of ethene with chlorine. The 1,2-dichloroethane that is formed undergoes thermal cracking to give chloroethene and hydrogen chloride.

$$C_2H_4Cl_2 \rightarrow C_2H_3Cl + HCl$$

The hydrogen chloride can then be converted to hydrochloric acid by passing it through water. A solution of high concentration can be produced easily because hydrogen chloride has a very high solubility in water. Hydrogen chloride gas is made up of covalent molecules – when dissolved in water it forms the hydrated ions $H^+(aq)$ and $Cl^-(aq)$.

You can learn about techniques for summarising information in **Activity ES5.2**, where you will be considering methods for manufacturing chemicals.

Assignment 8

a Use the concept of atom economy to explain why the production of hydrogen chloride by the direct combination of its elements is considered to be a good industrial method.

b A company is planning to set up a new chlor–alkali plant. They intend to use membrane cell technology to electrolyse brine for the production of chlorine, hydrogen and sodium hydroxide. They also plan to make hydrochloric acid by combining the hydrogen and chlorine. Write a brief report to summarise the factors they would need to consider when choosing their site under these headings:

   i   Access to raw materials
   ii   Transport links
   iii Waste disposal
   iv Safety
   v   Fixed costs
   vi Environmental issues.

ES6 *Treasures of the sea*

From simple chlorine-containing compounds such as methyl chloride, produced by volcanoes (e.g. the Uzon Caldera in Siberia, Russia) through to more complex molecules such as ephibatidine (a painkiller 500 times more potent than morphine) produced by the poison-arrow frog, the natural world produces a wide range of compounds containing halogens. To date, chemists have tabulated 2320 unique, naturally occurring organochlorine compounds, 2050 organobromine compounds, 115 organoiodines and 34 organofluorines. Bacteria, fungi, plants and animals – including humans – all produce compounds incorporating halogens.

Sponges, corals and most seaweed anchored to a reef share an inability to evade predators, and they provide defence mechanisms by synthesising halogen-containing compounds. The blue-green algae

▲ **Figure 23** The nudibranch *Chromodoris hamiltoni* (left) and blue-green algae *Lyngbya* (right) – both types of organism produce halogen-containing compounds as defence mechanisms.

Lyngbya synthesises aplysiatoxin, the cause of 'swimmers itch'. This is a condition that develops on parts of the body that have been exposed to water containing the algae. Reddened spots, called papules, form on the body within hours of exposure and itch intensely for several days before subsiding – one way of stopping swimmers invading water occupied by the algae!

Certain sponges produce brominated dioxins. Nudibranchs and sea hares, two varieties of shell-less saltwater slugs, also depend on organic compounds containing halogens. One species of sea hare synthesises and secretes a bitter-tasting, bromine-containing compound called panacene. This discourages predators, including sharks. Some nudibranchs lack the ability to synthesise such compounds and acquire them from their diet of sponges or algae.

Bromomethane is produced by seaweeds in the ocean and can react with chloride ions in the water to yield mixed chlorine- and bromine-substituted methanes such as chlorodibromomethane. Interconversions can take place where one halogen substitutes for another. These are often examples of **radical substitution reactions** – for example

$$2CH_3Cl + I_2 \rightarrow 2CH_3I + Cl_2$$

Alternatively, **hydrolysis** may lead to the loss of halogenoalkanes – for example

$$CH_3Cl + H_2O \rightarrow CH_3OH + HCl$$

Organohalogen compounds are often thought of as being solely industrial compounds, but many synthetic compounds are identical to those found naturally – indeed, even halogens are produced in nature. See Table 2 for a few examples.

Halogenoalkanes are organic compounds that consist of an alkane molecule with one or more hydrogen atoms replaced by halogen atoms. Some, like chloromethane and bromomethane, are produced naturally by reactions in the marine boundary layer. They are also released from vegetation and produced during forest fires.

During the twentieth century, a very wide range of halogenated alkanes were manufactured for a variety of uses including dry-cleaning, propellants for aerosols, refrigeration and as solvents. A realisation that many of these halogenated compounds, such as CFCs, were causing damage to our environment lead to worldwide agreements that stopped the production of many of these compounds and limited the range of uses of others. You can learn more about the impact of halogenoalkanes on the Earth's environment in **The Atmosphere**.

Chemical Ideas 13.1 tells you about how halogenoalkanes are named and the type of reactions they take part in.

You can learn more about how nucleophilic substitution reactions work in **Activity ES6.1**.

The ease with which halogenoalkanes can be made depends partly on which halogen atom, and the total number of halogen atoms, that are to be incorporated into the product molecule.

Activity ES6.2 considers the reactivity of the halogenoalkanes.

Reacting an alcohol with the appropriate hydrogen halide can make a monosubstituted halogenoalkane. For example, chloromethane can be made by reacting methanol with hydrogen chloride.

$$\underset{\begin{array}{c}|\\H\end{array}}{\overset{\begin{array}{c}H\\|\end{array}}{H-C-O-H}} + HCl \rightarrow \underset{\begin{array}{c}|\\H\end{array}}{\overset{\begin{array}{c}H\\|\end{array}}{H-C-Cl}} + H_2O$$

This reaction is an example of a **substitution reaction**. Most of the chloromethane manufactured industrially is made by this method. The reaction can be carried out either by bubbling hydrogen chloride gas through boiling methanol or by passing a mixture of methanol vapour and hydrogen chloride gas over a heated alumina catalyst.

You can make a halogenoalkane using **Activity ES6.3**.

A small amount of chloromethane is made commercially by reacting methane with chlorine. A mixture containing four chlorinated methane products forms as a result of this process. The four products

Table 2

Compound name	Formula	Synthesised industrially?	Found in nature?
tetrafluoroethene	C_2F_4	yes	yes
tetrachloromethane	CCl_4	yes	yes
dichloromethane	CH_2Cl_2	yes	yes
methyl chloride	CH_3Cl	yes	yes

contain different amounts of chlorine – with one, two, three or all four hydrogen atoms of the methane molecule having been substituted by a chlorine atom. The mixture is separated using fractional distillation.

One of the original uses of chloromethane, at the start of the twentieth century, was as a refrigerant but this stopped because of its high toxicity. Nowadays, the major use of chloromethane is in the manufacture of polymers.

▲ **Figure 24** In case a leak occurred during repair of a fridge, the engineer would wear a gas mask.

Some of the trichloromethane that forms in the reaction of methane with chlorine is used in the manufacture of PTFE. The trichloromethane is reacted with anhydrous hydrogen fluoride at 600 °C to form chlorodifluoromethane, $CHClF_2$. This was once produced in large quantities for use in refrigeration systems. However, as it is an ozone-depleting gas its use as a refrigerant is being phased out. You can find out more about ozone depletion in **The Atmosphere**. However, $CHClF_2$ is now being made in increasing quantities as a feedstock for the manufacture of polymers.

The uses that have been found for halogenoalkanes are linked to their reactivity. In some cases, uses have taken advantage of the lower reactivity of the fluorine- and chlorine-containing compounds. For example, PTFE has many uses that take advantage of its very low reactivity and the fact that it has a slippery, non-stick surface.

▲ **Figure 25** The roof of London's O_2 stadium is made from glass fibre coated with PTFE.

Safety is a major factor in the choice of method for the manufacture of a halogenoalkane. This is because the chemicals used to produce both the halogenoalkane and the product itself can be potentially harmful. The company needs to make sure that all processes are monitored to ensure that no leaks occur and methods need to be in place to deal with any waste products. It is also necessary to carry out the reactions under controlled conditions.

One of the problems with many organic syntheses is that the yield of the desired product is often low. In these situations, the scientists need to work out the best conditions for the process in order to maximise the yield.

You can find out how to calculate a percentage yield in **Chemical Ideas 15.7**.

Assignment 9

In industry, one method for making 1,2-dichloroethane is to react ethene with hydrogen chloride and oxygen:

$$CH_2=CH_2 + 2HCl + \tfrac{1}{2}O_2 \rightarrow CH_2ClCH_2Cl + H_2O$$

Under the optimum reaction conditions, 1.0 tonnes of ethene produces 3.36 tonnes of 1,2-dichloroethane.
a   Calculate the formula masses of ethene and 1,2-dichloroethane.
b   Calculate the maximum possible yield of 1,2-dichloroethane from 1.0 tonnes of ethene.
c   Calculate the percentage yield of 1,2-dichloroethane obtained in this process. Give your answer to 2 s.f.

ES7 *Summary*

In this module you have learned about the production of two elements, bromine and chlorine, from their raw materials. The processes involved rely on redox chemistry. Ideas about redox are also important in explaining many of the important reactions of the halogens. To understand the chemistry of the halogens you needed to know more about the energy levels in atoms and how electrons are arranged in these energy levels. The halogens show a trend in melting and boiling point as the group is descended. In order to explain this, you needed to know about the type of intermolecular bond that is present between molecules of the elements.

The processes by which bromine and chlorine are extracted, stored and transported involve hazards. In the cases of bromine and chlorine, these arise from the dangerous nature of the elements themselves.

Many of the bromine and chlorine compounds mentioned in this module have ionic structures, and much of their chemistry takes place in solution. You have had to learn more about ionic compounds and the processes of dissolving and precipitation, as well as how to calculate concentrations of substances in solution.

The halogens also form a range of covalently bonded compounds, including the halogenoalkanes. You have learnt about how they are made and the type of reactions they take part in.

Finally, the module gave you an insight into some aspects of the chemical industry – for example, the quantities of materials involved, the emphasis on safety, the importance of economic factors and environmental impact. You also learnt about how atom economy can influence the choice of method used to make a product and how to calculate the percentage yield of a process.

Activity ES7 will help you to check your notes on this module.

A

THE ATMOSPHERE

Why a module on 'The Atmosphere'?

The chemical and physical processes going on in the atmosphere have a profound influence on life on Earth. They involve a highly complex system of interrelated reactions, and yet much of the underlying chemistry is essentially simple. The focus of this module is change – change in the atmosphere brought about by human activities, and the potential effects on life. The influences of human activities in two major areas are explored – the depletion of the ozone layer in the upper atmosphere and the link between increased concentrations of greenhouse gases in the lower atmosphere and global warming.

Some important chemical principles are introduced and developed in considering these phenomena. In particular the effect of radiation on matter, the factors that affect the rate of a chemical reaction, the formation and reactions of radicals and the idea of dynamic equilibrium, as well as the specific chemistry of species that are met in the context of atmospheric chemistry – such as oxygen, carbon dioxide, methane and organic halogen compounds. The structure of carbon dioxide, CO_2, is compared to that of silica, SiO_2, which is found in many of the Earth's rocks.

Overview of chemical principles

In this module you will learn more about ideas you will probably have come across in your earlier studies:
- the effect of chlorofluorocarbons on the ozone layer
- rates of reactions
- the greenhouse effect
- reversible reactions.

You will learn more about some ideas introduced in earlier modules:
- the electromagnetic spectrum (**Elements of Life**)
- covalent bonding (**Elements of Life, Developing Fuels** and **Elements from the Sea**)
- electronegativity and bond polarity (**Elements from the Sea**)
- the chemistry of simple organic molecules (**Developing Fuels**)
- the use of moles and quantitative chemistry (**Elements of Life, Developing Fuels** and **Elements from the Sea**)
- enthalpy changes, enthalpy cycles and bond enthalpies (**Developing Fuels**)
- catalysis (**Developing Fuels**).

You will also learn new ideas about:
- the interaction of electromagnetic radiation with matter
- the formation and reactions of radicals
- factors that affect the rate of a reaction
- the nature of chemical equilibrium
- covalent network structures.

A

THE ATMOSPHERE

What's in the air?

The atmosphere is a relatively thin layer of gas extending about 100 km above the Earth's surface. If the world were a blown-up balloon, the rubber would be thick enough to contain nearly all the atmosphere. Thin though it is, this layer of gas has an enormous influence on the Earth. A simplified picture of the lower and middle parts of the atmosphere is shown in Figure 2. The two most chemically important regions are the troposphere and the stratosphere. Note the way that temperature changes with altitude. The atmosphere becomes less dense the higher you go. In fact, 90% of all the molecules in the atmosphere are in the troposphere. Mixing is easy in the troposphere because hot gases can rise and cold gases can fall. The reverse temperature gradient in the stratosphere means that mixing is much more difficult in the vertical direction. However, horizontal circulation is rapid in the stratosphere, particularly around circles of latitude.

▲ **Figure 1** The Earth as seen from space.

Table 1 (page 66) shows the average composition by volume of dry air from an unpolluted environment and is typical of the troposphere.

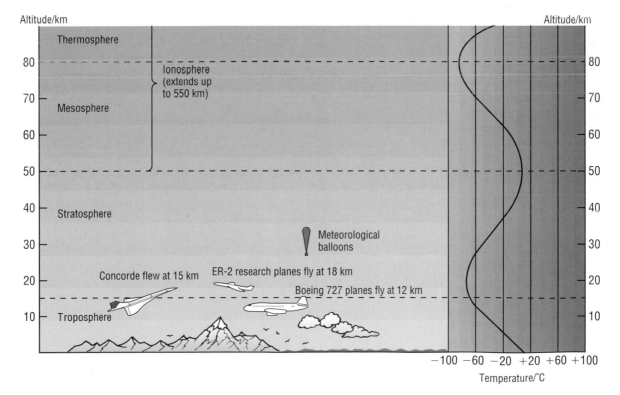

▲ **Figure 2** The structure of the atmosphere and the change in temperature with altitude.

Table 1 Composition by volume of dry tropospheric air from an unpolluted environment (* = variable).

Gas	Concentration (by volume)
	Per cent
nitrogen	78
oxygen	21
argon	1
	Parts per million
carbon dioxide	383*
neon	18.2
helium	5.2
methane	1.8*
krypton	1.1
hydrogen	0.5
dinitrogen oxide, N_2O	0.3*
carbon monoxide	0.1*
xenon	0.09
nitrogen monoxide, NO, and nitrogen dioxide, NO_2 (NO_x)	0.003*

The concentrations of some of these substances are measured in parts per million (ppm) by volume. This measure is often used when small concentrations are involved – 383 ppm corresponds to a percentage concentration of 0.0383%.

The atmosphere hasn't always had this composition. The first atmosphere was lost altogether during the upheavals in the early life of the solar system. The next atmosphere consisted of compounds such as carbon dioxide, methane and ammonia, which bubbled out of the Earth itself.

3000 million years ago there was very little oxygen in the atmosphere. But when the first simple plants appeared they began to produce oxygen through photosynthesis. For more than 1000 million years very little of this oxygen reached the atmosphere. It was used up as quickly as it was formed to oxidise sulfur and iron compounds and other chemicals in the Earth's crust. It wasn't until this process was largely complete that oxygen began to collect in the atmosphere.

THE FIRST PLANTS

The first plants appeared about 3000 million years ago and were probably single-celled organisms called cyanobacteria, or blue-green bacteria.

They lived in the surface waters of the oceans under anaerobic conditions, and used sunlight as their energy source to drive the chemical reactions needed to maintain their growth.

Similar organisms are still found on Earth today.

When the oxygen concentration reached about 10% there was enough for the first animals to evolve using oxygen for respiration. Eventually there was enough respiration and other processes going on to remove the oxygen as fast as it was formed. Since then, the oxygen concentration has remained at about 21%.

Look again at Table 1. All the gases listed are produced as a result of natural processes. Human activities add more gases to the atmosphere. Some of them, like carbon dioxide, are already present, but we increase their concentration. These gases are marked by an asterisk in Table 1, and their main sources as a result of human activities are shown in Table 2. Other gases in the atmosphere, like the chlorofluorocarbons (CFCs) and hydrofluorocarbons (HFCs), are produced *only* as a result of human activity.

Table 2 Sources of some of the gases in the atmosphere produced as a result of human activities.

Gas	Main source as a result of human activities
CO_2	combustion of hydrocarbon fuels (e.g. in power stations, motor vehicles); deforestation
CH_4	cattle farming; landfill sites; rice paddy fields; natural gas leakage
N_2O	fertilised soils; changes in land use (e.g. from the soil when land is ploughed up)
CO	incomplete combustion of hydrocarbons (e.g. from car exhausts)
NO and NO_2 (NO_x)	combustion of hydrocarbon fuels (from the reaction of N_2 and O_2)

Given time, gases always mix together completely and this natural diffusion process is greatly speeded up in the atmosphere by air currents and prevailing winds. So, in time, pollutant gases spread throughout the atmosphere. Atmospheric pollution is a global problem – it affects us all. In this module, we shall be looking at two global problems in particular:

- the depletion of the ozone layer in the stratosphere
- the link between increased concentrations of greenhouse gases in the troposphere and global warming.

Assignment 1

Use Table 1 to answer the following questions.
a   How many parts per million (by volume) of argon are there in a typical sample of tropospheric air?
b   In 1 dm³ of tropospheric air, what is
   i   the volume of methane present?
   ii   the percentage of methane *molecules* in the sample?

A2 *Screening the Sun*

The sunburn problem

▲ **Figure 3** The trend-setting clothes designer Coco Chanel, who set a new fashion for suntanned skin among white-skinned Europeans in the 1920s.

Until the 1920s a suntan was something a white-skinned person could not avoid if they had to work outdoors in the Sun. Those who didn't preferred to distinguish themselves by remaining pale. It was the clothes designer Coco Chanel who made sunbathing fashionable. She appeared with a golden tan after a cruise on the yacht belonging to the Duke of Westminster, who was one of the world's wealthiest men.

As we discover more about the effects of the Sun's radiation on the chemical bonds in living material, sunbathing seems less of a good idea.

The Sun radiates a wide spectrum of energy – part of this spectrum corresponds to the energy required to break chemical bonds. Sunlight can therefore break bonds, including those in molecules such as DNA in living material. This can cause damage to genes and lead to skin cancer. There appears to be a greater risk of skin melanomas occurring when pale skins are occasionally exposed to intense sunlight (for example, office workers sunbathing on holiday) than with the skins of people who work outdoors or who tan gradually.

On a less serious level, sunlight can damage the proteins of the connective tissue beneath the skin, so that years of exposure can make people look wrinkly and leathery. Even brief exposure to the Sun may cause irritation of the blood vessels in the skin, making it look red and sunburnt.

Many of the ideas in this storyline are linked to the interaction of radiation with matter. **Chemical Ideas 6.2** covers the different ways that radiation and matter can interact.

▲ **Figure 4** Over-exposure to sunlight causes severe sunburn and peeling skin and increases the risk of developing skin cancers.

Chemical sunscreens

Figure 5 (page 68) shows the effects of different parts of the Sun's spectrum on the skin.

You can see that the most damaging region of this spectrum is in the ultraviolet. Fortunately, there are chemicals which absorb much of this radiation.

Have you ever wondered why people don't get sunburnt indoors? You can sit by a window for hours on a sunny day without burning. The glass in the window lets through visible light but absorbs the damaging ultraviolet radiation, so it never reaches your skin. (Perspex does let through some ultraviolet, so it is possible to burn through Perspex.)

Chemists have found materials which do a similar job to glass that can be spread onto the skin. They are called *sunscreens* and tonnes of them are sold every summer.

But the best sunscreen of all is not made by chemists. It has always been with us – it is the atmosphere.

In **Activity A2.1** you can examine the effect on some substances of radiation from the Sun.

Activity A2.2 investigates the effectiveness of sunscreens.

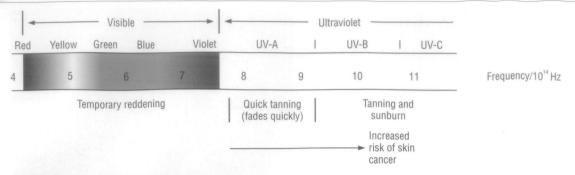

▲ **Figure 5** The effects of sunlight on the skin.

▲ **Figure 6** A sunscreen's ability to protect skin is expressed as a 'sun protection factor' or SPF number – SPFs indicate the time it will take for the Sun to produce a certain effect on your skin.

Why is the atmosphere such a good sunscreen?

Certain atmospheric gases absorb ultraviolet radiation strongly. They act as a global sunscreen, preventing much of the Sun's harmful radiation from reaching the Earth.

Most of this absorption goes on in the region of the upper atmosphere called the stratosphere (see Figure 2 on page 65). Particularly important is the gas **ozone**, which absorbs ultraviolet radiation in the region 10.1×10^{14}–14.0×10^{14} Hz. This is the region which is most damaging to your skin. (Look back at Figure 5, which shows the effect on skin of radiation of different frequencies.) Much of the damaging ultraviolet radiation is absorbed by ozone in the stratosphere.

There is no life in the stratosphere because the powerful ultraviolet radiation would break down the delicate molecules of living things. Indeed, even simple molecular substances get broken down in the stratosphere. Some of the covalent bonds break to give fragments of molecules (atoms or groups of atoms) called **radicals**.

Higher up in the atmosphere, above the stratosphere, the radiation is powerful enough to knock electrons out of the atoms, molecules and radicals. Ions are produced, which lead to the name of that part of the atmosphere – the *ionosphere*.

Assignment 2

Look back at Figure 2 on page 65, which shows the different regions of the atmosphere. Suggest the *types* of chemical particles – atoms, molecules, radicals and ions – which can be found in

a   the troposphere
b   the stratosphere
c   the ionosphere.

Activity A2.3 investigates the screening effect of different gases on solar radiation.

A3 *Ozone: a vital sunscreen*

Ozone is present in the atmosphere in only tiny amounts, dispersed among other atmospheric gases. If all the ozone in the atmosphere were collected and brought to the Earth's surface at atmospheric pressure, it would form a layer only 3 mm thick.

High up, in the stratosphere, ozone *protects* us by absorbing harmful ultraviolet radiation. However, lower down, in the troposphere, it can be a real nuisance if concentrations near ground level become too high (see the **Developing Fuels** storyline, **Section DF6**).

It isn't really surprising that there is so little ozone in the atmosphere – it reacts so quickly with other substances and gets destroyed. In fact, we might ask why the ozone in the atmosphere hasn't run out. Some reactions must be producing it too.

Many of the ideas in this part of the storyline are concerned with radicals. You can find out more about these by studying **Chemical Ideas 6.3**.

How is ozone formed in the atmosphere?

Ozone is formed when an oxygen atom (an example of a radical) reacts with a dioxygen molecule:

$$O + O_2 \rightarrow O_3$$
$$\text{ozone}$$

One way to make oxygen atoms is by splitting up (dissociating) dioxygen molecules. This requires quite a lot of energy – the bond enthalpy of the oxygen–oxygen bond in dioxygen is $+498\,kJ\,mol^{-1}$. In this case the energy can be provided by ultraviolet radiation, or by an electric discharge.

As soon as oxygen atoms have been produced, they react with the dioxygen molecules which are always present in the air. You can often smell the sharp odour of ozone near electric motors or photocopiers. The electric discharges happening inside the machine make some of the dioxygen molecules in the air dissociate into atoms. Have you ever noticed the ultraviolet lamps used to kill bacteria in food shops? You can often smell ozone near them too.

Some of the ozone in the troposphere is formed in the complex series of reactions taking place in photochemical smogs. These develop in bright sunlight over large cities which are heavily polluted by motor vehicle exhaust fumes (see the **Developing Fuels** storyline, **Section DF6**). In this case, oxygen atoms are produced by the action of sunlight on the pollutant gas nitrogen dioxide.

▲ **Figure 7** Helium-filled balloons are sent up into the stratosphere to measure ozone concentrations. A cord almost 10 miles in length attaches the balloon to measuring instruments on the launch vehicle – the cord can be reeled in and out to obtain measurements at different altitudes.

In the stratosphere, oxygen atoms are formed by the photodissociation of dioxygen molecules. This happens when dioxygen absorbs ultraviolet radiation of the right frequency.

The reaction can be summarised as:

$$O_2 + h\nu \rightarrow O + O \qquad \text{(reaction 1)}$$

In this reaction, $h\nu$ indicates the photon of ultraviolet radiation that is absorbed.

The oxygen atoms produced can do a number of things when they meet another chemical particle (an atom, radical or molecule) and collide with it. The least interesting of these is that the particles just bounce apart. But even when the oxygen atom collides with a particle that it can react with, not every collision results in a reaction.

More interesting is when the oxygen atom collides and reacts with an O_2 molecule, another O atom or an O_3 molecule. The three possible outcomes are represented by reactions 2–4 below:

$$O + O_2 \rightarrow O_3 \qquad \Delta H^{\ominus} = 106\,kJ\,mol^{-1} \quad \text{(reaction 2)}$$
$$O + O \rightarrow O_2 \qquad \Delta H^{\ominus} = 498\,kJ\,mol^{-1} \quad \text{(reaction 3)}$$
$$O + O_3 \rightarrow O_2 + O_2 \quad \Delta H^{\ominus} = 392\,kJ\,mol^{-1} \quad \text{(reaction 4)}$$

Reaction 2 is, of course, the one that produces ozone.

When the ozone absorbs radiation in the $10.1 \times 10^{14} - 14.0 \times 10^{14}\,Hz$ region, some molecules undergo photodissociation and split up again:

$$O_3 + h\nu \rightarrow O_2 + O \qquad \text{(reaction 5)}$$

It is this reaction which is responsible for the vital screening effect of ozone, since it absorbs the radiation responsible for sunburn.

Photodissociation and the subsequent reactions of the radicals produced are investigated in **Activities A3.1** and **A3.2**. You cannot use the damaging radiation needed to break down O_2 or O_3, so in the activities you will be working with Br_2 which absorbs light in the visible/near ultraviolet region.

Ozone – here today and gone tomorrow

You can see from reactions 1–5 on page 69 that ozone is being made and destroyed all the time. Left to themselves, these reactions would reach a point where ozone was being made as fast as it was being used up:

> rate of producing ozone = rate of destroying ozone

At this point, the concentration of ozone would remain constant. This is called a *steady state*.

It's like the situation in Figure 8 when you are running water into a basin with the plug out of the waste pipe. Before long, you get to the point where water is running out as fast as it's running in, and the level of water in the basin stays constant. If you turned the tap on more, the level of the water would rise – but that would make the water run out faster because of the higher pressure. Before long you would get to a steady state again, but this time with more water in the basin. What would happen if you made the waste pipe larger?

▲ **Figure 8** One example of a steady state situation.

The water coming out of the tap is like the reactions producing ozone, and the water going down the waste pipe is like the reactions that destroy it.

To estimate the concentration of ozone in the stratosphere, you need to know the **rates** of the reactions that produce and destroy ozone.

Chemists have studied these reactions in the laboratory and can write mathematical equations giving the rates of all the reactions involved in producing and destroying ozone. So, for example, taking reactions 1–5 on page 69, the rate of producing ozone will be the

rate of reaction 2. If there is a steady state, this will be balanced by the rate of destruction of ozone – the rate of reaction 4 *plus* that of reaction 5. From these relations, chemists can work out what the concentration of ozone *should* be at different altitudes, at different times of day and at different times during the year.

However, when chemists compare their *calculated* concentrations of ozone with *measured* values, they find that the actual concentration of ozone is a good deal *lower* than expected.

This suggests that the ozone is being removed faster than expected. Going back to the analogy of the basin and the running tap, it's as if the waste pipe had been made larger. But by what?

Assignment 5

A number of factors can affect the rate of ozone production and destruction. Remembering that the reactions are taking place in the gas phase and in strong sunlight, suggest *three* factors that could affect the rates of these reactions.

Explain how altering each factor would affect the rate of reaction.

You can investigate the effects of concentration and temperature on the rate of a chemical reaction in **Activity A3.3**.

The factors that affect the rate of a chemical reaction are described in **Chemical Ideas 10.1**.

The effect of temperature on the rate of a chemical reaction is discussed in more detail in **Chemical Ideas 10.2**.

What is removing the ozone?

We have seen that ozone is very reactive and reacts with oxygen atoms. But oxygen atoms aren't the only radicals to be found in the stratosphere. There are other radicals which can remove ozone by reacting with it.

Two important examples of radicals that react in this way are the chlorine atom (Cl) and the bromine atom (Br). Small amounts of chloromethane, CH_3Cl, and bromomethane, CH_3Br, reach the stratosphere as a result of natural processes (see Figure 9). Once in the stratosphere, their molecules are split up by solar radiation to give chlorine atoms and bromine atoms, respectively.

Other chlorine-containing compounds reach the stratosphere, in greater concentrations, as a result of human activities. These also absorb high energy solar radiation and break down to give chlorine atoms.

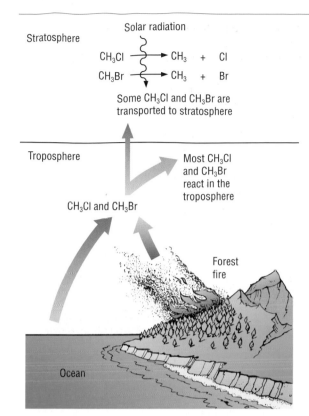

▲ **Figure 9** Chloromethane (CH_3Cl) and bromomethane (CH_3Br) are given off from the oceans and from burning vegetation. They are responsible for the small amounts of naturally produced chlorine and bromine in the stratosphere. (CH_3Br is also released into the atmosphere as a result of human activity. It is used as a fumigant for soils and grain.)

Chlorine atoms react with ozone like this:

$$Cl + O_3 \rightarrow ClO + O_2 \qquad \text{(reaction 6)}$$

and bromine atoms react in the same way.

The ClO formed is another reactive radical and can react with oxygen atoms:

$$ClO + O \rightarrow Cl + O_2 \qquad \text{(reaction 7)}$$

So now we have two reactions competing with each other to remove ozone from the stratosphere:

$$O + O_3 \rightarrow O_2 + O_2 \qquad \text{(reaction 4)}$$

and

$$Cl + O_3 \rightarrow ClO + O_2 \qquad \text{(reaction 6)}$$

The concentration of Cl atoms in the stratosphere is much less than the concentration of O atoms. So how significant is reaction 6?

It is in situations like this that it is very important for chemists to know something about the *rates* at which reactions occur.

The reaction of O_3 with Cl atoms would not matter much if it took place a lot more slowly than the reaction of O_3 with O atoms.

Chemists measured the rates of these two reactions in the laboratory under different conditions. They showed that, at temperatures and pressures similar to those in the stratosphere, the reaction of O_3 with Cl atoms takes place more than 1500 times *faster* than the reaction of O_3 with O atoms.

Even when they took into account the fact that Cl atoms have a much lower concentration than O atoms in the stratosphere, the chemists still found that the reaction with Cl atoms takes place sufficiently quickly to make a very large contribution to the removal of ozone.

What's more, the Cl atoms used in reaction 6 are regenerated in reaction 7 – and can then go on to react with more O_3. Adding together the equations for reactions 6 and 7 gives the equation for the overall reaction.

$$Cl + O_3 \rightarrow ClO + O_2 \qquad \text{(reaction 6)}$$

and

$$ClO + O \rightarrow Cl + O_2 \qquad \text{(reaction 7)}$$

Overall reaction: $O + O_3 \rightarrow O_2 + O_2$

The chlorine atoms act as a **catalyst** for this reaction. By going through the catalytic cycle many times, a single chlorine atom can remove about 1 million ozone molecules. So you can see why even low concentrations of Cl atoms can be devastating.

You can write a similar catalytic cycle involving bromine atoms. In fact, although the concentration of bromine atoms is much lower than that of chlorine atoms, bromine is about 100 times more effective in destroying ozone than chlorine!

You can read more about the role of catalysts in **Chemical Ideas 10.6**.

You can investigate the effect of a catalyst on a chemical reaction in **Activity A3.4**.

Other ways ozone is removed

Chlorine and bromine atoms aren't the only radicals present in the stratosphere which can destroy ozone in a catalytic cycle in this way.

If we represent the radical by the general symbol X, we can rewrite reactions 6 and 7 as a general catalytic cycle:

$$X + O_3 \rightarrow XO + O_2$$

and

$$XO + O \rightarrow X + O_2$$

Overall reaction: $O + O_3 \rightarrow O_2 + O_2$

Two other important radicals (HO, the hydroxyl radical, and NO, nitrogen monoxide) which can destroy ozone in this way are described on page 72.

COMPETING REACTIONS

Activation enthalpies are very important when *comparing* the rates of two competing reactions taking place under similar conditions. If the activation enthalpy of a reaction is large, only a small proportion of the colliding particles will have enough energy to react, so the reaction proceeds slowly. However, if the activation enthalpy is very small then most of the colliding particles will have sufficient energy to react and the reaction occurs very quickly.

Figure 10 shows the activation enthalpies for the reactions of O atoms and Cl atoms with ozone (O_3). The reaction with the lower activation enthalpy (i.e. Cl + O_3) will proceed more quickly.

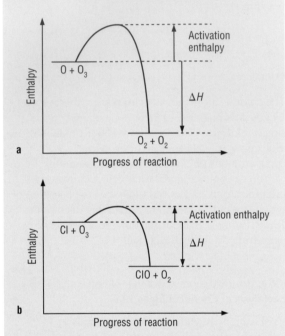

▲ **Figure 10** Plots of enthalpy changes as the reactant particles come closer together and react to form products.

Hydroxyl radicals (HO)
These are formed by the reaction of oxygen atoms with water in the stratosphere. They react with ozone like this:

$$HO + O_3 \rightarrow HO_2 + O_2$$

The HO_2 radicals then go on to react with oxygen atoms to re-form the HO radicals:

$$HO_2 + O \rightarrow HO + O_2$$

So this is another example of a catalytic cycle, and the HO radicals released can go on to react with more O_3 molecules.

▲ **Figure 11** George Porter shared the 1967 Nobel Prize for Chemistry for his work on very fast reactions using a technique called **flash photolysis**. Nowadays, a brief intense flash from a laser starts the reaction. The composition of the mixture is measured spectroscopically with a carefully timed second flash. Reactions that take place in nanoseconds (1 ns = 1×10^{-9} s), or even picoseconds (1 ps = 1×10^{-12} s) and femtoseconds (1 fs = 1×10^{-15} s), can be studied in this way.

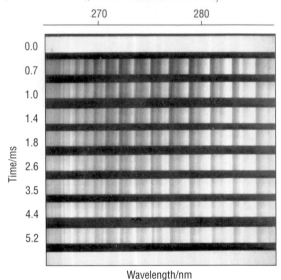

The series of spectra above were taken in the early days of the technique and were first published in the Royal Society of Chemistry journal *Discussions of the Faraday Society* in 1953. They were taken at different times following the flash photolysis of a mixture of chlorine and oxygen. The dark vertical lines indicate the absorption of radiation by the radical. Initially there is no ClO present, but its concentration rises very rapidly after the flash and then decays as it reacts.

Assignment 6

a Write an equation to show the formation of HO radicals from O atoms and water.

b Write equations to show how nitrogen monoxide can destroy ozone in a catalytic cycle.

Nitrogen monoxide (NO)
Nitrogen monoxide reacts with ozone to form nitrogen dioxide (NO_2) and dioxygen. Nitrogen dioxide can then react with oxygen atoms to release nitrogen monoxide and dioxygen to complete the catalytic cycle.

NITROGEN MONOXIDE IN THE STRATOSPHERE – WHERE DOES IT COME FROM?

Most of the nitrogen monoxide (NO) in the stratosphere is formed naturally. But an increasing proportion is produced as a result of the reaction of dinitrogen oxide (N_2O) with oxygen atoms:

$$N_2O + O \rightarrow 2NO$$

Dinitrogen oxide is the most abundant oxide of nitrogen in the troposphere (see Tables 1 and 2 on page 66). It is produced by biological reactions on Earth and some is carried up into the stratosphere. Dinitrogen oxide is released by bacteria which break down nitrogen compounds in the soil and also by bacteria in the oceans. But not all of it comes from natural processes. The increased use of fertilisers contributes to increasing dinitrogen oxide concentrations.

NO and NO_2 are both radicals. They are unusual radicals because they are relatively stable molecules and they can be prepared and collected like ordinary molecular substances. (It is important to remember that not all radicals are highly reactive.)
The radicals mentioned in this section (Cl, Br, HO and NO) are important, but they are only part of the whole picture. Hundreds of reactions have been suggested which affect the gases in the stratosphere.

Many of these have been going on since long before there were humans on Earth. But human activities can have a serious effect on certain key reactions, and so lead to dramatic changes in the concentration of ozone in the stratosphere.

Activity A3.5 will help you to check your understanding of the information that has been presented in **Sections A1** to **A3**.

A4   *The CFC story*

Chemists' concerns that substances put into the atmosphere by human activities may be destroying the ozone layer were raised in the early 1970s in connection with high-flying jet aircraft. Jet engines release nitrogen oxides (mostly NO) in their exhaust gases – could this make a significant difference to the amount of NO_x ($NO + NO_2$) in the stratosphere and so damage the ozone layer? In the end, it turned out that this wasn't a significant problem because the number of aircraft concerned was then too small to make much difference.

Alarm bells were rung in 1974 when Sherry Rowland published a paper predicting that chlorofluorocarbons (CFCs) would damage the ozone layer if they reached the stratosphere.

SHERRY ROWLAND'S PREDICTIONS

▲ **Figure 12** Professor Sherwood ('Sherry') Rowland.

Professor Sherry Rowland is the American scientist who, with Professor Mario Molina, predicted back in the early 1970s that chlorofluorocarbons (CFCs) would damage the ozone layer. He described to us how he made his discovery.

'I originally started looking at the CFC compounds as an interesting *chemical* problem in an environmental setting: whether we could predict from our laboratory knowledge what the fate of the CFCs would be in the Earth's atmosphere. At the beginning we were not thinking of the destruction of the CFCs as an environmental *problem*, but rather as just something that would be happening there.

'I've always been attracted to chemical problems – my first research experiment after graduate school involved putting a powdered mixture of lithium carbonate and ordinary glucose in the neutron flux of a nuclear reactor. A nuclear reaction in lithium produces tritium, a radioactive isotope of hydrogen, and I wanted to see if the tritium atoms could replace hydrogen atoms in glucose. (They did – and this led to many further interesting experiments in a field called *hot atom* chemistry.)

'With CFCs, we wanted to find out how quickly they would break down in the atmosphere, and by what chemical process. I was working with a young post-doctoral research associate called Mario Molina – this was in 1973 at the University of California at Irvine – and we knew that CFCs are very stable compounds. But how stable? How long could they hang around in the atmosphere? A year or a hundred years?

'Mario and I looked at all of the processes that could conceivably affect CFCs in the troposphere, and calculated how rapidly such reactions could

▶

occur. The answer was very slowly indeed. CFCs remain unreacted for *many decades* – even centuries.

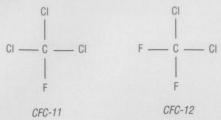

CFC-11   CFC-12

'So now we knew that CFCs survive unchanged for a very long time. But we also knew that when they eventually reach the stratosphere, they must be broken down by the fierce ultraviolet radiation there – everything is! The CFCs contain atoms of chlorine, fluorine and carbon and their ultraviolet breakdown releases free chlorine atoms. For example:

$$CCl_3F \rightarrow CCl_2F + Cl$$

'So we did some calculations to find out how many chlorine atoms would be formed now and in the future, and then asked what would happen to those chlorine atoms. In the stratosphere, we found that chlorine atoms are about a thousand times more likely to react with ozone than with anything else, leaving still another chlorine-containing chemical, chlorine oxide (ClO).

'So, once more we asked the same kind of question: what was going to happen to chlorine oxide in the stratosphere? And we found that it would react with oxygen atoms, releasing atomic chlorine again. The two reactions seem to go around in circles – chlorine atoms form chlorine oxide; chlorine oxide forms chlorine atoms – a seemingly endless chain. But with ozone being destroyed at every step! We then calculated how much ozone could be destroyed – each chlorine atom on average destroys about 100 000 molecules of ozone, and mankind has been putting about 1 million tonnes of CFCs into the atmosphere every year since the 1970s.

'We couldn't believe the answer! The calculated ozone loss was so high that we thought we must have moved a decimal point by mistake! But we checked very carefully and couldn't find any errors – there really was that much chlorine up there, and with much more expected in the future the ozone losses would eventually be enormous!

'We had started on this problem at the beginning of October 1973, and by mid-December we realised we were onto something very important. When a scientist makes a discovery, the first instinct is to publish it so that other scientists can learn about it and test the ideas with their own experiments. But the calculated ozone loss was so large that we wanted to be extra certain that no mistake had been made.

'So we visited Hal Johnson, at the University of California at Berkeley, because he had played a major role in showing how nitrogen oxides from high-flying aircraft could affect stratospheric ozone. From Hal we learned that the chlorine chain reaction with ozone had just been discovered, but without a source for chlorine. And we had discovered that the CFCs would be an enormous source of chlorine released directly into the stratosphere!

'Now, we were sure that we had discovered something really significant – and it was a major environmental problem. We published our results and conclusions in the scientific journal *Nature* and waited to see if others could pick holes in them. Many tried – because that's the way science works – but none succeeded. The issue of CFCs and stratospheric ozone had really arrived.

'The state of Oregon banned CFCs as propellant gases in aerosols in 1975, and the whole of the US, as well as Canada, Norway and Sweden, followed in the next 2 or 3 years. Unfortunately, the rest of Europe and Japan did not ban aerosol propellant use and the other major applications – refrigeration, insulation, cleaning electronics, etc. – continued to grow until the appearance of the hole in the Antarctic ozone layer resulted in international action in 1988–1990.'

For their work in atmospheric chemistry, Professors Sherry Rowland and Mario Molina, together with Professor Paul Crutzen, were awarded the Nobel Prize for Chemistry in 1995.

As the module develops you will see how their worries have been heeded.

The predictions come true

Science is all about making predictions, then testing them experimentally. The problem with Sherry Rowland's predictions was that they involved a long time-scale. What is more, they needed a large laboratory to test them. In fact they needed the largest laboratory in the world – the Earth's atmosphere.

But in 1985, scientists examining the atmosphere above the Antarctic made a momentous discovery.

Since the mid-1980s, monitoring of ozone concentrations has continued with even greater urgency. Satellite readings and measurements from balloons and high-altitude planes have supported the measurements taken from the ground, and have confirmed the presence of the 'hole' over the Antarctic.

Large losses of total ozone in Antarctica reveal seasonal ClO$_x$/NO$_x$ interaction

J. C. Farman, B. G. Gardiner & J. D. Shanklin

British Antarctic Survey, Natural Environment Research Council, High Cross, Madingley Road, Cambridge CB3 0ET, UK

Recent attempts[1,2] to consolidate assessments of the effect of human activities on stratospheric ozone (O$_3$) using one-dimensional models for 30°N have suggested that perturbations of total O$_3$ will remain small for at least the next decade. Results

▲ **Figure 13** The headline-making paper which appeared in the scientific journal *Nature* in May 1985, reporting the discovery of a 'hole' in the ozone layer over Antarctica.

JOE FARMAN'S STORY

▲ **Figure 14** Dr Joe Farman.

Dr Joe Farman is the British scientist whose group first discovered the 'hole' in the ozone layer. We went to Cambridge to talk to him, and this is how he described their discovery.

'To stand up and make a fuss you need to have your background secure.

'We were measuring ozone concentrations over the Antarctic as part of our work with the British Antarctic Survey. We use ultraviolet spectroscopy – ozone absorbs ultraviolet radiation of a particular frequency. If you measure how strongly the atmosphere is absorbing ultraviolet of that frequency, you can work out the concentration of ozone in that part of the atmosphere. Our measurements started in 1981, and over 2 or 3 years I became convinced that there was something seriously wrong. The concentrations of ozone were much lower than expected, particularly in October, which is the Antarctic spring. In 1984 we put in a new instrument to check our readings and that gave us confirmation. There really was a 'hole' in the ozone layer.

'We were making our measurements from *below*, by looking up into the atmosphere from the Earth's surface at our base in Antarctica. The Americans were making similar measurements, but from *above* – NASA satellites were using ultraviolet spectroscopy to measure ozone concentrations by looking down.

'So why didn't NASA spot the 'hole' first? The trouble is that NASA satellites make *too many* measurements. They collect enormous quantities of data on all sorts of things, not just ozone concentrations.

'They can't possibly process all the data, so their computers are programmed to ignore any data that seem impossibly inaccurate. In the case of the ozone measurements this meant that they ignored most of them because they were so far out from what was expected. About 80% of the data for October were discarded by the computer because no-one believed that ozone concentrations could get that low.

'Later, when our own measurements showed the concentration of ozone really was that low, NASA went back and re-examined the discarded data, which confirmed our own measurements.

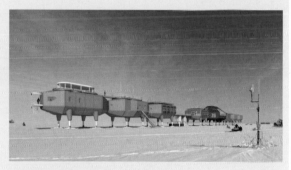

▲ **Figure 15** A computer-generated image of the new building to be constructed in 2007 at the British Antarctic Survey's Halley Base. Each module is held above the snow surface on legs that can be raised or lowered. The legs are mounted on large skis, so the station can be towed by bulldozers to a different site.

'In 1985 we published our findings in the scientific journal *Nature*. Then it was a matter of convincing the world how serious the problem is. Once a scientist has made a momentous discovery of this kind, there is a duty to tell everyone about it.

'To me the real horror is the sheer speed with which it has happened. In 1985 the United Nations published a report saying there was plenty of time to study the ozone problem. At about the same time we published our own results, which show that the ozone gets turned over in a period of about *5 weeks*! We still don't know exactly what the effects of ozone depletion will be, but when ▶

you are affecting a system that fast you have to be aware that almost anything could happen.

'We may think that we understand a lot about the way that nature works, but to my mind we are so ignorant that *anything* could happen. However good your models are you have to *keep making measurements* to make sure that nature really does work the way you think.

'It's easy to see now, with the benefit of hindsight, where we went wrong with CFCs. The original idea was to look for something *very stable* which would not be flammable, poisonous or corrosive. But the trouble with very stable substances is that they stay around for a long time – and when they do eventually break down they form something *very unstable* and reactive, which is likely to cause trouble.'

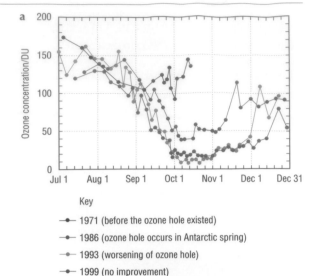

Key

—•— 1971 (before the ozone hole existed)

—•— 1986 (ozone hole occurs in Antarctic spring)

—•— 1993 (worsening of ozone hole)

—•— 1999 (no improvement)

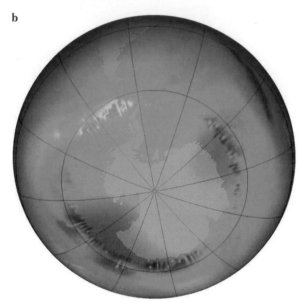

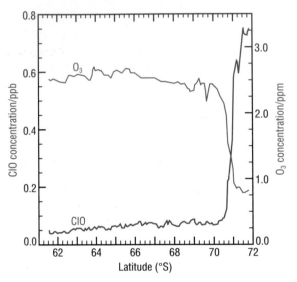

▲ **Figure 16** ER-2 aircraft fly through the stratosphere loaded with scientific instruments. These graphs show measurements of ClO radicals (in parts per billion) and ozone (in parts per million) recorded at 18 km altitude. (Note that the concentrations of the two species are different by a factor of 10^3.) The measurements provided convincing evidence that Cl radicals are involved in ozone depletion – scientists had found the 'smoking gun'.

▲ **Figure 17**　**a** The graph shows how the ozone concentration varied during the year over the Amundsen–Scott South Pole Station for four years between 1971 and 1999.　**b** The ozone hole over the Antarctic in September 2006 was the most severe ever recorded for area and depth. Maps like this are obtained every day by satellite and can be seen on the Internet.

CFCs: very handy compounds

In 1930 the American engineer Thomas Midgley demonstrated a new refrigerant to the American Chemical Society. He inhaled a lungful of dichlorodifluoromethane (CCl_2F_2) and used it to blow out a candle.

Midgley was flamboyantly demonstrating two important properties of CCl_2F_2 – its lack of toxicity and its lack of flammability. Up to that time, ammonia had been the main refrigerant in use. Ammonia has a

convenient boiling point, $-33\,°C$, which means it can easily be liquefied by compression. Unfortunately it is also toxic and very smelly, and often caused leakage problems.

Midgely had been asked to find a safe replacement for ammonia. For a substance to be a good refrigerant, it has to be easily compressible to perform well in the cycle of compression and evaporation. It needs a boiling point low enough to evaporate efficiently, and a freezing point low enough for the refrigerant not to freeze in the refrigerator. In addition, it should be odourless, non-toxic, chemically stable and non-flammable. CCl_2F_2 passed all the tests and was seen as the ideal solution. Refrigerators became cheaper and more efficient, and were soon widely available.

CCl_2F_2 belongs to a family of compounds called chlorofluorocarbons (CFCs), which contain chlorine, fluorine and carbon. There are several members of the family, all with different boiling points. That's one of the things that made them so useful – there were CFCs with boiling points to suit different applications. CFCs were used:

- as refrigerants and in air conditioning units
- as propellants in aerosol cans (for example, for deodorant spray and insect repellents)
- as blowing agents for expanded plastics such as polystyrene and polyurethane used in insulating foam
- as solvents in dry cleaning and to degrease electronic circuits.

By the early 1970s, industry was producing about a million tonnes of CFCs a year.

◀ **Figure 18** CCl_2F_2 was widely used as the refrigerant in food refrigerators and in air-conditioning units. Eventually some of the refrigerant ended up in the atmosphere, through leakage and when the refrigerator was scrapped.

Could the trouble with CFCs have been foreseen?

When Thomas Midgley and other chemists developed CFCs, they did their job too well. They found a family of compounds that are very unreactive, and this makes them excellent for the jobs listed above. The trouble is that they are *too* unreactive.

Scientists remained unaware that CFCs could affect the atmosphere for over 40 years. In hindsight, it is easy to apportion blame. You have to look back at how the science of the atmosphere developed to understand why this was so.

At the time when CFCs were being tested, questions about environmental consequences for the atmosphere were simply not on the agenda. Atmospheric chemistry was in its infancy and the stratosphere seemed too far away to be of concern. Sensitive instruments that could detect minute concentrations of compounds in the air did not exist.

In 1970, the UK scientist James Lovelock built an electron capture detector which could measure small concentrations, and used it to show that CFCs had been transported to remote locations. Even then, because of the stability of CFCs, they were thought to be harmless. The work of Sherry Rowland and Mario Molina raised concerns in 1974.

As new techniques were developed, instruments were fitted in satellites and high-altitude aircraft and were taken up by balloons to monitor the concentration of chemical substances in the stratosphere.

The reactions taking place in the atmosphere were studied under laboratory conditions, and computer modelling techniques allowed scientists to simulate the processes occurring and so make predictions. In the 1980s and 1990s all this contributed to an explosion of knowledge of the chemistry of the atmosphere, and this work continues today.

It is now known that most organic compounds released into the atmosphere are broken down in the troposphere by chemical scavengers (such as HO radicals) and never reach the stratosphere. This doesn't happen with CFCs – their stability, initially seen as a huge advantage, proved to be their downfall.

Chlorine reservoirs

Chlorine atoms are obviously bad news in the stratosphere – they attack ozone and destroy it. Indeed, they are so reactive that, left to themselves, they would quickly destroy most of the ozone there. Fortunately, they have not done that yet because there are other molecules in the stratosphere that react with Cl atoms.

Methane, CH_4, is an important example of such a molecule. It is produced on Earth in large quantities by living organisms, as you will see later in **Section A6**. Most of the methane released is oxidised in the troposphere, but significant amounts of it are eventually carried up into the stratosphere.

Once in the stratosphere, methane molecules remove chlorine atoms by reacting with them like this:

$$CH_4 + Cl \rightarrow CH_3 + HCl$$

Hydrogen chloride, HCl, is called a *chlorine reservoir molecule* because it stores chlorine in the stratosphere. Another important reaction which produces a chlorine reservoir molecule is

$$NO_2 + ClO \rightarrow ClONO_2$$

The product is chlorine nitrate, $ClONO_2$.

Some of these reservoir molecules may eventually be carried down into the troposphere. Both are soluble in water and will be removed in raindrops. Most, however, remain in the stratosphere with serious consequences, as you will see next.

Why does the ozone hole develop over the poles?

Satellite measurements have shown that ozone concentrations have decreased in other parts of the globe too, but the effect is particularly dramatic in the Antarctic spring (see Figure 17b, page 76).

There are two reasons for this and both are associated with the special weather conditions occurring in Antarctic. Firstly, the very low temperatures (below $-80\,°C$) that occur in the polar winter when the Sun disappears lead to the formation of *polar stratospheric clouds*. These clouds are made up of tiny solid particles – some are mainly particles of ice, others are rich in nitric acid, HNO_3. These particles provide surfaces on which chemical reactions can occur.

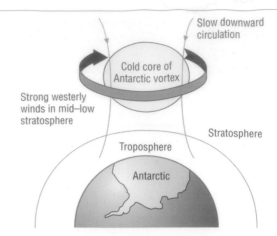

▲ **Figure 20** The winter vortex over the Antarctic. The cold core is almost isolated from the rest of the atmosphere.

▲ **Figure 19** Polar stratospheric clouds over the Antarctic. Years in which the temperature is low, with extensive formation of polar stratospheric clouds, correspond to particularly wide and deep ozone holes.

Secondly, during the Antarctic winter a vortex of circulating air forms, which effectively isolates the air at the centre of the vortex. This cold core turns into a giant sealed reaction vessel (Figure 20).

The chlorine reservoir molecules – HCl and chlorine nitrate, $ClONO_2$ – are **adsorbed** onto the surface of the solid particles in the polar stratospheric clouds, where they react together:

$$ClONO_2 + HCl \rightarrow Cl_2 + HNO_3$$

The HNO_3 remains dissolved in the ice particle, but the Cl_2 molecules are released as a gas trapped in the isolated core of the vortex. In the Antarctic spring, when sunlight returns, the vortex starts to break up and the Cl_2 molecules break down to form Cl atoms.

$$Cl_2 + h\nu \rightarrow Cl + Cl$$

This results in the dramatic loss of ozone in the early Antarctic spring. These reactions are summarised in Figure 21.

Although ozone depletion is most severe in the Antarctic, ozone depletion also occurs over the Arctic. The effect is more variable here because temperatures are not usually as low as those above the Antarctic. Polar stratospheric clouds are less abundant and less persistent and a stable polar vortex does not form. However, ozone depletion here extends over more densely populated areas across northern latitudes, including Canada, the US and Europe, and is a cause for concern.

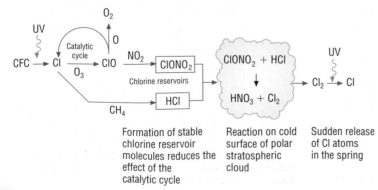

◄ **Figure 21** How reactions in polar stratospheric clouds contribute to the dramatic loss of ozone in the Antarctic spring.

Assignment 7

Here are three reactions involving chlorine atoms from the stratosphere:

$$CH_4 + Cl \rightarrow CH_3 + HCl \qquad \text{(reaction 8)}$$
$$Cl + Cl \rightarrow Cl_2 \qquad \text{(reaction 9)}$$
$$Cl + Cl_2 \rightarrow Cl_2 + Cl \qquad \text{(reaction 10)}$$

a For each of these reactions, answer the following questions.

   i Which bond, if any, is broken during the reaction? Use the **Data Sheets** to find the bond enthalpy of this bond.

   ii What new bond is made during the reaction? Use the **Data Sheets** to find the bond enthalpy of this bond.

   iii Use the bond enthalpies to find a value for ΔH for the reaction.

b Which of the reactions have the effect of removing chlorine atoms from the stratosphere?

c The energy required to get the bond-breaking/bond-forming processes going in a reaction is called the *activation enthalpy*. Explain why reaction 8 has a much higher activation enthalpy than reaction 9.

d Despite its higher activation enthalpy, reaction 8 normally removes Cl atoms more rapidly than reaction 9. Can you suggest a reason why?

In **Activity A4** you can practise your extended writing skills by summarising and explaining the ways in which scientists established the link between CFCs and ozone depletion.

A5 *What is the state of the ozone layer now?*

Ozone is a vital sunscreen gas which protects us from ultraviolet radiation. It was clear in the 1980s that removing it from the stratosphere would have serious consequences for the Earth. The trouble was that we did not know exactly what the full extent of these would be.

One thing was certain – the numbers of cases of skin cancers and eye cataracts were increasing as the ozone was destroyed. It has been estimated that reducing ozone by 10% could cause a 30–50% increase in skin cancer cases.

But what about species other than humans? Increased ultraviolet radiation could affect species such as plankton in the oceans. That in turn could affect other organisms involved in the food chain.

And what about the weather? Changes in the amount of radiation reaching the Earth will affect the temperature of the Earth itself, which of course affects the weather.

Success for global cooperation

Governments realised that it was not worth risking the global experiment needed to find the answers to these questions. In 1987, at an international meeting in Montreal, a procedure was agreed for restricting the production and release of CFCs into the atmosphere. The agreement covered not only CFCs but related bromine-containing compounds, such as $CBrClF_2$. These are called *halons* and were used in fire-fighting systems.

Since then, new global observations of ozone depletion have prompted a series of amendments to strengthen the treaty. The aim is to reduce and eventually eliminate emissions of ozone-depleting substances as a result of human activity. The list of these now includes bromomethane and other halogenoalkanes.

By 1998 the developed nations had almost phased out their use of CFCs, though they are still used for some specialised appliances such as asthma inhalers. A special fund was set up to help developing countries to move away from CFC use by 2010.

▲ **Figure 22** A report written to celebrate the twentieth anniversary of the Montreal Protocol summarises its achievements over the last 20 years.

World unites on ozone deal

Paul Brown
Environment Correspondent

A worldwide agreement to phase out CFCs and other ozone depleting chemicals by the year 2000 was reached in London last night.

A group of environmentally advanced countries pushed to have the date brought forward to 1997, but were blocked by the United States, the Soviet Union, and Japan. The compromise solution was a 50 per cent reduction by 1995, 85 per cent by 1997, and 100 per cent by 2000. A subclause agreed the position would be reviewed in 1992, to see if the timetable could be improved.

The agreement provided for the establishment of a new global front to help the Third World adapt to the changes.

Chris Patten, the Environment Secretary, said: "This is a major step forward in environmental diplomacy. It is a unique agreement bringing together, as it does, the establishment of environmental objectives with provision of funds and the transfer of technology."

It bodes well for future global environmental agreements.

Answering criticisms that the agreement still did not go far enough, he said: "We would all like to have stopped CFCs production tomorrow but this was the best agreement that was possible, taking all considerations into account".

Negotiations had run well over the time yesterday as detailed timetables for phasing out chemicals were hammered out for inclusion in the agreement.

There was personal success for Mr Patten, who, as chairman of the conference, redrafted the final document so that both China and India felt they could pledge to sign the Montreal Protocol.

By yesterday, 59 nations had signed and most of the other 39 at the conference were expected to ratify soon.

In spite of ministerial joy at the agreement, there were doubts that the timetable would be quick enough to prevent ozone depletion being a serious problem.

Yesterday's deal meant that CFCs, the main ozone depleter, used in fridges and air conditioning, will go by 2000. Methyl chloroform, a metal cleaning agent, will be banned by 2005.

Much of the argument centred on how quickly individual chemicals could be phased out.

One of the triumphs was getting India and China, with more than one third of the world's population between them, to join. Maneka Gandhi, the Indian Environment Minister, had held out for two days for the transfer of technology from the West to be included in the agreement so India could manufacture CFCs substitutes itself.

Mrs Gandhi said yesterday the agreement now said that if technology was not transferred, that India did not have to stop the manufacture of CFCs. That placed the onus on the West to keep its promises, and on that basis she was prepared to recommend her government to ratify the protocol.

Joe Farman, the British scientist who discovered the hole in the ozone layer, said he feared the agreement was still not stringent enough.

Mr Farman, head of atmospheric dynamics at the British Antarctic Survey, said the ozone hole would go on getting bigger for some time.

Chlorine is currently 3.6 parts per billion in the atmosphere. This would grow to 4.8 parts per billion in 10 years, he said. He calculated this could lead to 18 per cent depletion in ozone in the northern hemisphere during the winter and spring by the year 2000.

Under the agreement, he calculated that it would be 2030 before the chlorine level went down to the 1986 level when the hole was first announced.

▲ **Figure 23** *The Guardian* newspaper reporting, in June 1990, on the international agreements reached in London about reducing CFCs. Even so, the effect of CFCs on the ozone layer will increase for many years yet.

CFC replacements

The chemical industry moved very rapidly to find replacements for CFCs that would have no significant damaging effect on the ozone layer. In the short term, replacement compounds were hydrochlorofluorocarbons (HCFCs), such as $CHClF_2$. The difference is that the molecules contain $H-C$ bonds and HCFCs are broken down in the troposphere before they have time to reach the stratosphere. The first step in the breakdown is the reaction with HO radicals:

$$CHClF_2 + HO \rightarrow CClF_2 + H_2O$$

The $CClF_2$ radical formed then goes on to react further.

Unfortunately, some molecules of $CHClF_2$ will make it to the stratosphere where they will photodissociate to release Cl atoms. HCFCs will be phased out in developed countries by 2020 and in the rest of the world by 2040.

For the longer term, hydrofluorocarbons (HFCs), such as CH_3CF_3, are seen a better option because they have no ozone-depleting effect, even if they make it to the stratosphere.

Table 3 compares the properties and uses of some CFCs and related compounds with those of HCFCs and HFCs.

Sadly, there is no perfect solution. Both CFCs and their current replacements are greenhouse gases and contribute to **global warming** (a problem you will meet later in this module in **Section A6**). There has also been some concern about the fate of HFCs in the troposphere, since their degradation products include HF and trifluoroethanoic acid, CF_3COOH, but it is felt that the concentrations are too small to be a problem.

When will the ozone layer recover?

The concentrations of CFCs in the troposphere peaked around the beginning of this century and are now starting to decrease. Figure 24 shows the results from monitoring air samples at different sites around the world. As you would expect, the concentrations of HCFCs and HFCs, though lower than those of the CFCs, are rising.

CFCs are very stable and their long lifetimes in the atmosphere mean that recovery of the ozone layer will be slow – and will be interrupted by year to year variations caused by stratospheric weather conditions and solar activity. For example, it seemed in 2002 that the ozone layer was starting to recover, but this turned out to be a false hope. The ozone depletion over Antarctica in September 2006, shown in Figure 17b, was the most severe to date. Scientists predict that the ozone hole will start to recover around 2010, but complete recovery is not expected until 2060–2070.

Table 3 Properties and uses of some CFCs and related compounds compared with those of HCFCs and HFCs. The ozone-depleting potential (ODP) is a measure of the effectiveness of the compound in destroying stratospheric ozone. CFC-11 is defined as having an ODP of 1.0.

Compound	Code	ODP	Lifetime/years	Uses
Main culprits				
CCl_3F	CFC-11	1.0	45	refrigeration, air conditioning,
CCl_2F_2	CFC-12	1.0	100	foams, cleaning solvents
$CBrF_3$	Halon-1301	10.0	65	firefighting
$CBrClF_2$	Halon-1211	3.0	16	
CCl_4	–	1.1	26	solvents
CH_3CCl_3	–	0.1	5	
Short-term solutions				
$CHClF_2$	HCFC-22	0.06	12	replaces CFC-11 and CFC-12
$CHCl_2CF_3$	HCFC-123	0.02	1	replaces CFC-11
Long-term solution				
CH_3CF_3	HFC-143a	0	52	replaces CFC-11 and CFC-12

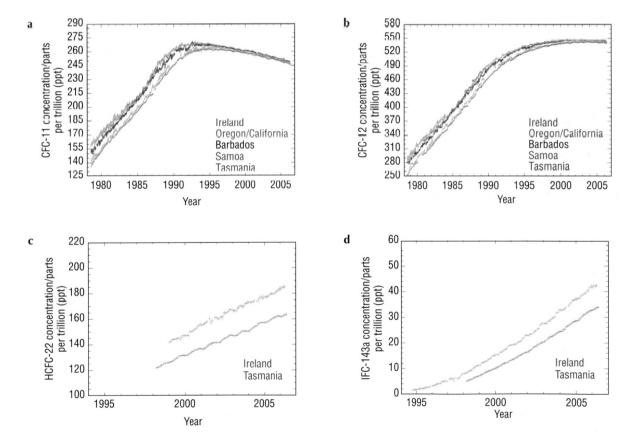

▲ **Figure 24** The results of monitoring air samples at different sites around the world
a CFC-11,   **b** CFC-12,   **c** HCFC-22,   **d** HFC-143a.

CELEBRATING A 20-YEAR REUNION

Scientists from the British Antarctic Survey met in 2005 to commemorate their discovery of the ozone hole 20 years earlier.

Jonathan Shanklin said:

'The 2005 hole is larger and deeper than the holes that formed when the discovery was made, but the situation would be much worse if the Montreal Protocol had not come into force. This agreement shows us that global action by governments to stop the release of ozone-depleting chemicals really can help society to successfully mitigate a global environmental problem.

'We are still experiencing large losses of Antarctic ozone each spring because CFCs and other chemicals live for a long time in our atmosphere. However, the ban ensures that we will see an improvement in the future. We now need to take similar actions to control greenhouse gases – otherwise we will bequeath future generations a significantly different climate from that of today.'

▲ **Figure 25** Joe Farman, Brian Gardiner and Jonathan Shanklin – the team that discovered the hole in 1985 – with the Dobson ozone spectrophotometer that they used to measure stratospheric ozone concentrations.

▲ **Figure 26** Children in New Zealand have to wear hats on school outings in summer to protect them from harmful ultraviolet radiation.

A6 *The greenhouse effect*

This story now moves a little nearer to the Earth – to the troposphere, the bottom 15 km or so of the atmosphere. But for the moment we will stick with methane, which in the stratosphere helps remove the chlorine atoms which destroy the ozone layer.

In the troposphere, methane's role has a less helpful side to it. To see why, we need to look at the way the Sun keeps the Earth warm.

Radiation in, radiation out

When things get hot, they send out electromagnetic radiation – the hotter the object, the higher the energy of the radiation.

The surface of the Sun has a temperature of about 6000 K and this means that it radiates energy in the ultraviolet, visible and infrared regions.

The Earth is heated by the Sun's radiation. The Earth's average surface temperature is about 285 K (12 °C) – a lot cooler than the Sun, but still hot enough to radiate electromagnetic radiation. At this lower temperature, the energy radiated is mainly in the infrared region. The situation is illustrated in Figure 27.

The radiation from the Sun that reaches the outer limits of the atmosphere is mainly in the visible and ultraviolet regions. Part of this energy is absorbed by the Earth and its atmosphere, and part is reflected back

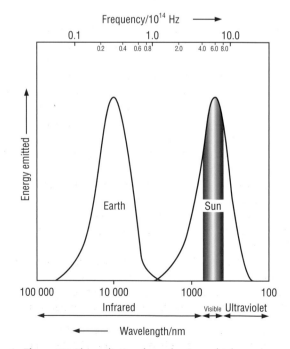

▲ **Figure 27** The radiation from the Sun which reaches the outer limits of the atmosphere, and the radiation given off from the surface of the Earth. (The frequencies and wavelengths are plotted on a logarithmic scale, so each division is a factor of 10 greater than the one before.)

HOW IS METHANE FORMED?

As humans, we are used to living in an airy world. We use aerobic respiration to oxidise carbohydrates such as glucose to carbon dioxide and water:

$$C_6H_{12}O_6 + 6O_2 \rightarrow 6CO_2 + 6H_2O$$
glucose

But some organisms live in airless places, under anaerobic conditions. Instead of turning carbohydrates to CO_2 and water, they turn them to other less oxidised materials. Yeast is a good example – it converts glucose to ethanol and CO_2.

An important group of bacteria – called *methanogenic bacteria* – work in anaerobic conditions to turn materials such as carbohydrates to methane and other related substances.

Methanogenic bacteria are very common. In fact, you can be pretty sure that wherever carbohydrate is left in anaerobic conditions, methanogenic bacteria will be present and methane will be produced. Since most biological material contains carbohydrate (or substances that can be converted to carbohydrate), it follows that methane is produced whenever biological material decays anaerobically. This may occur in:

- marshes, compost heaps and land-fill sites, where vegetation rots without air
- rice paddy fields, where water and mud cut off air from rotting vegetation (Figure 28)

- the vast stretches of tundra in Arctic regions when the ice melts and the ground becomes waterlogged
- the digestive tracts of animals, where part-digested food is acted on by bacteria (a cow releases about $500\,000\,cm^3$ of methane every day in belches!).

Methane is one of the most abundant of the trace gases in the troposphere (see Table 1 on page 66). Its concentration in the troposphere is now about 2.5 times what is was in pre-industrial times.

▲ **Figure 28** Rice paddy fields cover large areas of land and are one of the biggest sources of methane.

into space. The part that gets absorbed helps to heat the Earth, and the Earth in turn radiates energy back into space. A steady state is reached where the Earth is radiating energy as fast as it absorbs it. Under such conditions, illustrated by Figure 30 (page 85), the average temperature of the Earth remains constant.

As in all steady states, the delicate balance can be disturbed by changes to the system – in particular by changes to the quantities of various gases in the atmosphere. Methane is one example.

Methane is an example of a **greenhouse gas**. It absorbs some of the infrared radiation emitted from the surface of the Earth and prevents it being re-radiated into space. The effect of this is to make the Earth warmer.

To understand how absorption of infrared radiation by methane molecules causes warming, we need to think about what happens to the energy once it is absorbed by the molecules. Two things can happen:

- some infrared radiation is re-emitted by the molecules – but this happens *in all* directions so some is radiated back towards the Earth, and some is radiated out towards space;
- absorption of infrared radiation increases the vibrational energy of the methane molecules and

the bonds vibrate more vigorously. This vibrational energy can be transferred to other molecules in the air (such as N_2 and O_2 molecules) by collisions. This increases their kinetic energy – they move faster – raising the temperature of the air.

The overall effect is to trap some of the radiation from the Earth that would otherwise be lost.

WHY IS IT CALLED THE GREENHOUSE EFFECT?

The warming of the Earth by gases such as methane was called the **greenhouse effect** because it was thought to be similar to the way in which the inside of a greenhouse is kept warmer than the air outside. This is illustrated in Figure 29, with an attempt to show the relative wavelengths of the radiation involved.

In this case, the glass is acting like the greenhouse gases in the atmosphere. The glass is transparent and allows the incoming visible radiation from the Sun to pass through, but it absorbs the infrared radiation emitted by the soil and plants in the greenhouse. The glass re-emits some of this radiation, but this happens *in all directions* so some is radiated back into the greenhouse.

However, the analogy is not a good one because the main way that a greenhouse works is by preventing the warm air escaping by convection and as a result of air currents – rather than preventing radiation escaping. Even so, the term *greenhouse effect* is widely used, so it helps to understand the original analogy.

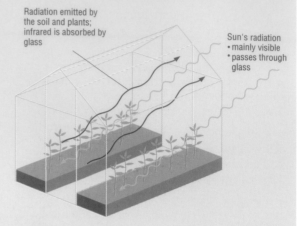

Radiation emitted by the soil and plants; infrared is absorbed by glass

Sun's radiation
• mainly visible
• passes through glass

▲ **Figure 29** One of the ways in which a greenhouse keeps plants warm.

Do other gases have a greenhouse effect?

Methane is not the only gas to behave in this way. Carbon dioxide and several other gases in the troposphere absorb infrared radiation, but not the visible or ultraviolet light from the Sun. This means that they are greenhouse gases. They let the Sun's visible radiation in, but they stop some of the Earth's infrared radiation getting out – they contribute to the greenhouse effect, which makes the Earth warmer than it would otherwise be.

Activity A6.1 looks at the absorption characteristics of some atmospheric gases.

Figure 30 shows the balance between incoming and outgoing energy for the Earth.

Some gases have a more powerful greenhouse effect than others. A number of factors must be taken into account when deciding how much each gas contributes to the overall greenhouse effect. These include how efficient the gas is at absorbing infrared radiation, the concentration of the gas and its average lifetime in the troposphere.

Scientists have developed a number of indicators to assess these factors. One useful indicator is the **global warming potential**, which depends on the absorption efficiency of the gas and its atmospheric lifetime. It is used to compare the effects caused by gases over a given time, relative to the same mass of carbon dioxide – carbon dioxide is assigned a value of 1.

Table 4 lists the global warming potentials of some gases, along with their abundances and average lifetimes in the troposphere.

The global warming potentials in Table 4 compare the effects, over a 100-year period, of releasing 1 kg of the gas and 1 kg of carbon dioxide into the atmosphere. Using the data in the table you can calculate that 1 molecule of methane, for example, has the same effect as about 10 molecules of carbon dioxide.

Assignment 9

You can get a rough idea of the relative contribution of a gas to the total greenhouse effect in the atmosphere by multiplying the global warming potential by the abundance of the gas.

a Use the data in Table 4 to list the gases in order of how much they contribute to the total greenhouse effect.

b Which of these gases are produced in significant amounts by human activities?

c Which gas in part **b** would it be most fruitful to try to control in order to control the greenhouse effect?

The greenhouse effect is good for you

Without the greenhouse effect we would not be here. By trapping some of the Sun's radiation, the atmosphere keeps the average temperature of the Earth high enough to support life.

If we had no atmosphere on Earth, it would be like the Moon – barren and lifeless. The surface of the

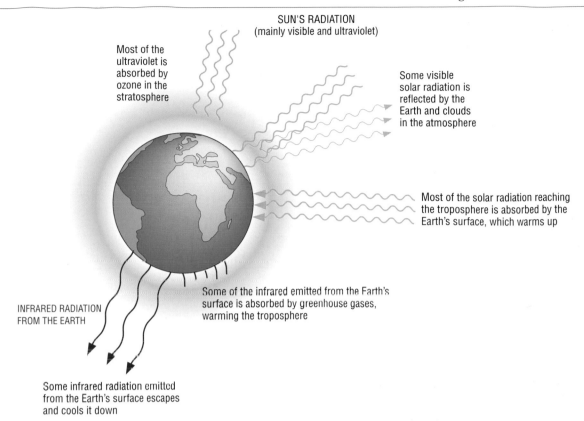

SUN'S RADIATION
(mainly visible and ultraviolet)

Most of the ultraviolet is absorbed by ozone in the stratosphere

Some visible solar radiation is reflected by the Earth and clouds in the atmosphere

Most of the solar radiation reaching the troposphere is absorbed by the Earth's surface, which warms up

INFRARED RADIATION FROM THE EARTH

Some of the infrared emitted from the Earth's surface is absorbed by greenhouse gases, warming the troposphere

Some infrared radiation emitted from the Earth's surface escapes and cools it down

▲ **Figure 30** The Earth – input and output of energy.

Table 4 Relative contributions to the greenhouse effect of various gases in the atmosphere. (The global warming potentials are calculated for a 100 year period.)

Gas	Tropospheric abundance by volume (%)	Global warming potential	Average lifetime in the troposphere/years
N_2	78	negligible	
O_2	21	negligible	
Ar	1	negligible	
CO_2	3.8×10^{-2}	1	150
CH_4	1.8×10^{-4}	25	12
N_2O	3.2×10^{-5}	298	114
CCl_3F (CFC-11)	2.5×10^{-8}	4750	45
CCl_2F_2 (CFC-12)	5.4×10^{-8}	10 900	100
CH_3CClF_2 (HCFC-142b)	1.5×10^{-9}	2400	19

Moon gets very hot during the day, but is bitterly cold at night.

If, on the other hand, the composition of our atmosphere resembled that of our neighbouring planet Venus, with 97% carbon dioxide, the greenhouse effect would make it so hot that life-forms would find it impossible to survive.

Look at Figure 31 (page 86), which compares the conditions on the surfaces of Earth, Venus and Mars. The carbon dioxide atmosphere on Venus is thick and the greenhouse effect is extreme. Our other neighbouring planet, Mars, has an atmosphere which is mostly carbon dioxide, like that on Venus. But on Mars the atmosphere is very thin. There is only a small greenhouse effect and so Mars is cold.

We are used to the stable, comfortable temperature on Earth and small changes in that temperature could have a dramatic effect on life.

In **Activity A6.2** you can simulate the greenhouse effect in the laboratory.

Approximate composition (%)	Carbon dioxide: 97	Nitrogen: 78	Carbon dioxide: 95
	Nitrogen: 3	Oxygen: 21	Nitrogen: 3
Average surface temperature/°C	464	15	−60
Estimated warming due to greenhouse effect/°C	500	40	3

▲ **Figure 31** A comparison of the atmospheres and surface temperatures of Earth and its neighbouring planets. The estimated warming due to the greenhouse effect is obtained by working out the difference between the observed temperature on the planet and the temperature expected just taking into account its distance from the Sun.

A7 Trouble in the troposphere – what happens if concentrations of greenhouse gases increase?

The temperature of the Earth is determined by the balance of radiation absorbed by the atmosphere and the surface of the Earth and the radiation emitted into space, as shown in Figure 30.

About 100 years ago, the Swedish chemist Arrhenius predicted that increasing concentrations of carbon dioxide could lead to warming of the Earth. Average temperatures did indeed rise from 1880 to 1940, by about 0.25 °C. But then between 1940 and 1970 they fell again, by 0.2 °C.

So why are we worried today? During the 1970s, measurements of carbon dioxide in the atmosphere began to show a significant increase (see Figure 32) and new predictions began to be made about the effect of the carbon dioxide increase on the Earth's climate.

Making predictions about the climate is very difficult because it involves so many variable factors, some of which are still poorly understood.

The information available is used to construct *computer models* of the Earth's climate. The amount of data fed into the models is enormous. It includes, for example, information about emissions of all the gases into the atmosphere – both from natural sources and as a result of human activities – and how these emissions are expected to change in the future. This is combined with mathematical descriptions of all the physical and chemical processes occurring in the atmosphere and the interactions of the atmosphere with the oceans and the surface of the land – plus other

factors such as the variations in the intensity of the Sun's radiation. So you can see why powerful computers are needed.

The models are used to describe and predict the global climate, as well as regional weather patterns, and are constantly being improved as more reliable information becomes available.

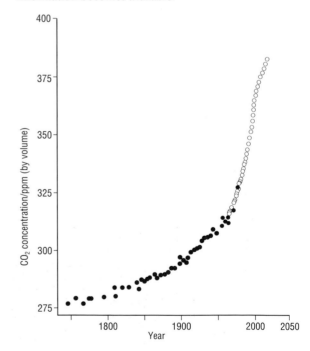

▲ **Figure 32** Carbon dioxide concentrations in the atmosphere from 1750. The data before 1958 were obtained by analysing the ice deep below the surface of Antarctica, which had trapped carbon dioxide year by year (closed circles). The open circles show data from the air, obtained at Mauna Loa (see Figure 38, page 90).

In 1988, the World Meteorological Organization and the United Nations set up the Intergovernmental Panel on Climate Change (IPCC) to assess the problem of potential global warming and climate changes that might occur as a result.

The panel produces regular reports that represent the work of thousands of scientists from around the world. The second IPCC report in 1995 led to the adoption of the Kyoto Protocol in 1997, in which a total of 169 countries agreed to the proposed limits on emission of greenhouse gases. The agreement came into effect in 2004.

Since 1988, the IPCC has collected a huge body of scientific evidence to support the link between increased concentrations of greenhouse gases in the atmosphere and the warming of the Earth.

Figure 33 shows differences in average near-surface temperatures over the period 1850 to 2006. The differences are relative to the average global temperature between the years 1961 and 1990, which is taken as the baseline. Scientists calculate an average temperature for any particular year from measurements from around the world, and calculate the difference between the temperature that year and the 1961–1990 average.

The records of surface temperatures from around the world going back 150 years suggest that the 11 years in the period 1995–2006 rank among the 12 warmest years on record.

One of the methods that climate scientists employ to test their models is to use them to predict past climates – so that they can compare the accuracy of their predictions with what actually happened.

Figure 34 shows two sets of simulations of global

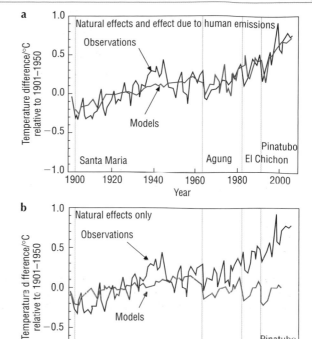

Figure 34 Model simulations of average global temperature differences relative to the period 1901–1950.
a The models here include emissions of greenhouse gases due to human activity as well as natural effects. The blue line shows the mean of several models.
b The models here include only natural effects. The blue line shows the mean of several models.
In both graphs, the vertical grey lines indicate the timing of major volcanic eruptions, which result in short-term cooling, and the red lines show the observed temperatures for comparison.

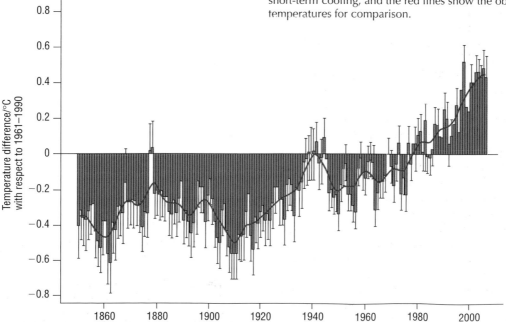

▲ **Figure 33** Observed changes in global near-surface temperatures over the period 1850–2006. The red bars relate to the global temperature for each year. The blue line shows the average trend.

temperature differences (this time from a 1901–1950 baseline). In the first graph, the models include warming from the greenhouse gases in the atmosphere due to human emissions. In the second graph, the models include only natural effects, such as solar energy changes and volcanic emissions. In both graphs the red lines show the observed temperature.

As a result of modelling studies like those in Figure 34, the IPCC panel concluded that it is 'extremely unlikely' (a probability of <5%) that the global pattern of warming over the last 50 years can be explained without including warming due to human emissions. Scientists believe that the period of cooling between 1940 and 1970 was caused by other climate factors, which masked the warming created by increased concentrations of carbon dioxide.

▲ **Figure 35** The eruption of Mt Pinatubo in the Philippines in June 1991 released enormous clouds of dust and sulfur dioxide, which led to some global cooling over the next few years.

So far we haven't mentioned the role of water vapour in the troposphere. The average concentration of water vapour is about 1%, though it varies considerably from place to place.

Like carbon dioxide, water vapour absorbs infrared radiation and is a greenhouse gas. Together, carbon dioxide and water vapour are the most significant greenhouse gases. Because they are so abundant in the atmosphere, they absorb a lot of the infrared radiated by the Earth. Water vapour makes the larger contribution – simply because more of it is present.

Water, however, is different from other greenhouse gases. Under most conditions on Earth, water is a liquid with some vapour associated with it. $H_2O(l)$ and $H_2O(g)$ are quickly interconverted. If human activities, such as burning fuels, put $H_2O(g)$ into the atmosphere, most of it will condense to $H_2O(l)$ and eventually return to Earth. So, although a lot of water vapour is released into the atmosphere as a result of human activities, its concentration depends on temperature and is controlled naturally by the processes of evaporation and condensation in the water cycle. In that sense, $H_2O(g)$ isn't nearly as much of a greenhouse problem as $CO_2(g)$.

But there are two other things to consider. First, as the Earth gets warmer more $H_2O(g)$ will evaporate from the oceans. The warmer air will be able to hold more water vapour and this will increase global warming – and so on, enhancing the original warming effect. This is an example of *positive feedback*.

Second, the droplets of liquid water in clouds tend to block out the Sun, as people living in the UK know well. Depending on the balance between these two effects, more water in the atmosphere *could* work in either direction – to increase or to decrease overall warming. This is one reason why climate modellers find it so hard to predict exactly what will happen to the Earth's climate in the future.

As Figure 36 shows, carbon dioxide and water absorb in two bands across the Earth's radiation spectrum. Between these two bands is a 'window' where infrared radiation can escape without being absorbed. In fact, about 70% of the Earth's radiation escapes into space through this 'window'.

The role of other greenhouse gases

Gases produced by human activities can increase the natural greenhouse effect of the atmosphere. There are two types:

- Gases already naturally present in the atmosphere, which are increased in amount by human activities. Carbon dioxide is an important example. Humans

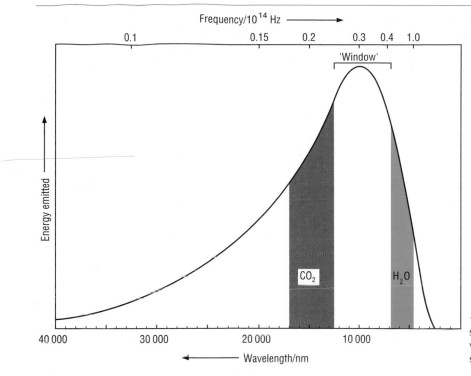

◀ **Figure 36** The Earth's radiation spectrum, showing the regions where CO_2 and H_2O absorb strongly.

▲ **Figure 37** It is difficult to predict accurately the effect of atmospheric water on global warming. Water is a greenhouse gas, but water droplets in low clouds tend to block out the Sun and cool the surface of the Earth.

burn fossil fuels and this increases the amount of carbon dioxide in the atmosphere. This in turn increases the greenhouse effect.
* Gases that are not naturally present in the atmosphere. These gases may absorb radiation in the vital 'window' through which radiation normally escapes into space. CFCs are an example. Although CFCs are present in the atmosphere in only small amounts, they are important because they have a very large global warming potential – so each molecule has a big effect.

Ozone is another example of a greenhouse gas that absorbs radiation in the 'window'. Ozone is not emitted directly into the atmosphere, so it is not assigned a global warming potential. Nevertheless, it does contribute to the overall warming. Ozone is a *secondary pollutant* (see **Developing Fuels, Section DF6**) formed in the troposphere by photochemical reactions involving NO_x and volatile organic compounds, such as those given off in vehicle exhausts. It is reactions such as these that contribute towards the formation of photochemical smog.

A9 *Focus on carbon dioxide*

At least half of the expected increase in the greenhouse effect due to human activities is likely to be caused by carbon dioxide. So control of the greenhouse effect must focus mainly on control of the amount of carbon dioxide we produce.

Measuring carbon dioxide

Because carbon dioxide is so crucial in the greenhouse effect, it is important to be able to make accurate measurements of its concentration in the atmosphere. Then we can keep an eye on how its concentration is changing.

The proportion of carbon dioxide in the atmosphere is fairly small – about 0.038%. In your earlier science work you probably used the limewater test to detect carbon dioxide. This is fine when you want a *qualitative* test and when the concentration of carbon dioxide is fairly large, but it's not nearly sensitive enough to measure the small changes in atmospheric carbon dioxide concentration that occur from year to year. What is needed is a sensitive *quantitative* method for measuring carbon dioxide concentration.

The method most commonly used by researchers is *infrared spectroscopy*. Carbon dioxide absorbs infrared radiation – that's the reason for all the trouble. The more carbon dioxide there is, the stronger the absorption.

Using infrared measurements, it is possible to get an accurate picture of the way carbon dioxide concentration has changed over the years. The graph in Figure 38 shows measurements of atmospheric carbon dioxide concentrations made at Mauna Loa Observatory, situated at an altitude of about 3500 m in Hawaii.

▲ **Figure 39** The Mauna Loa Observatory, Hawaii.

Assignment 10

a The graph in Figure 38 shows a zig-zag pattern. What times of the year correspond to
  i the peaks?
  ii the troughs?
b Suggest a reason for this zig-zag pattern.
c What was the average percentage increase in carbon dioxide concentration at Mauna Loa between 1960 and 2007?
d What problems would be involved in making a similar record of carbon dioxide concentration in London?

Where does the carbon dioxide come from – and go to?

The increase in concentration of carbon dioxide in the Earth's atmosphere is due to the increasing use of fossil fuels. Over the last 100 years, the amount of fossil fuels burned has increased by about 4% every year. Because of a fall in demand in the former Soviet Union, the consumption of fossil fuels levelled off during the early 1990s but is now increasing again, partly due to rapid economic growth in emerging industrial nations like China and India.

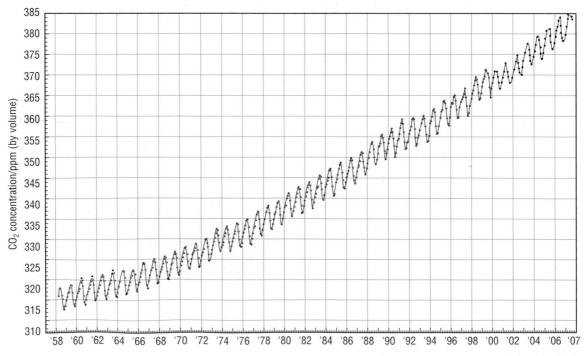

▲ **Figure 38** The build up of CO_2 in the atmosphere as recorded at Mauna Loa Observatory, Hawaii.

Assignment 11

In 2001, the concentration of carbon dioxide in the atmosphere at Mauna Loa (see Figure 38) was 370 ppm (by volume). The total mass of carbon dioxide in the atmosphere was estimated to be about 2760 Gt at that time. (1 Gt = 1 gigatonne = 1×10^9 tonnes.)

Through the four years from 2001 to the end of 2004, it is estimated that 34 Gt of fossil fuels were burned throughout the world.

In 2004 the measured percentage of carbon dioxide in the atmosphere at Mauna Loa was found to be 377 ppm (by volume).

Now answer these questions:

a Assume that the 34 Gt of fossil fuels burned in the four years were all carbon (this is a reasonable approximation since the other main element in fossil fuels, hydrogen, has a very low relative atomic mass). Calculate the mass of carbon dioxide formed by burning this mass of fuel.

b Calculate the new total mass of carbon dioxide in the atmosphere at the end of 2004, assuming that the extra carbon dioxide came only from burning fossil fuels and that it stayed in the atmosphere.

c By what percentage would you expect the concentration of carbon dioxide in the atmosphere to have increased in the four years?

d By what percentage did the concentration of carbon dioxide in the atmosphere actually increase in the four years?

e Comment on your answers to parts **c** and **d**.

Calculations of the kind you have just done suggest that – judging by the quantity of fossil fuels being burned – the increase in carbon dioxide concentration in the atmosphere *should* have been almost *twice* as much as it actually has been. Atmospheric carbon dioxide concentration is not increasing as fast as we might expect. Good news – but where is all the carbon dioxide going?

Oceans soak up carbon dioxide

Oceans cover almost three-quarters of the Earth's surface. Carbon dioxide is fairly soluble in water, so large amounts of atmospheric $CO_2(g)$ dissolve in the oceans.

To understand the chemistry in this section you need to know about dynamic equilibrium. **Chemical Ideas 7.1** will help you with this.

When carbon dioxide dissolves in water, it forms hydrated CO_2 molecules:

$$CO_2(g) + aq \rightleftharpoons CO_2(aq)$$

(reaction 11)

This is a reversible reaction, as you will know if you have watched what happens when you take the top off a bottle of fizzy drink. However, it is a fairly slow reaction (which is a good thing from the point of view of fizzy drink consumers) and it takes quite a long time for equilibrium to be reached.

The uptake of carbon dioxide by the oceans is quicker than this. Minute marine plants called phytoplankton use up most of the carbon dioxide that goes into the sea (Figure 40). So the concentration of 'free' $CO_2(aq)$ is small and gaseous carbon dioxide is encouraged to dissolve.

A small proportion of the $CO_2(aq)$ (about 0.4%) goes on to react with the water. It forms hydrogencarbonate ions (HCO_3^-) and hydrogen ions:

$$CO_2(aq) + H_2O(l) \rightleftharpoons HCO_3^-(aq) + H^+(aq)$$

(reaction 12)

Since $H^+(aq)$ ions are formed, this reaction is responsible for the acidic nature of carbon dioxide. A solution of carbon dioxide in pure water is, however, only weakly acidic. Reaction 12 does not go to completion – an equilibrium is set up with products and reactants both present in solution. Only 0.4% of the $CO_2(aq)$ reacts, so there is very much more $CO_2(aq)$ than $H^+(aq)$ present.

You can investigate some chemical equilibria in **Activity A9.1**.

Assignment 12

a What will happen to the position of equilibrium in reaction 11 if the concentration of atmospheric carbon dioxide increases?

b What will be the effect on reaction 12 of the change described in part **a**?

c In Assignment 11 you found that the proportion of atmospheric carbon dioxide was not rising as fast as expected. Suggest a reason for this.

d Do you think the carbon dioxide in the oceans and in the atmosphere are actually in equilibrium? Explain your answer.

▲ **Figure 40** Phytoplankton act as a 'biological pump', removing CO_2 from the atmosphere and transporting organic carbon compounds from surface waters to deeper layers as a 'rain' of dead and decaying organisms. This is balanced by upward transport of carbon by deeper water, which is richer in CO_2 than surface water.

Increased uptake of carbon dioxide by the oceans changes its chemical equilibrium and the oceans become more acidic. Measurements over the last 20 years show an average decrease of about 0.02 pH units per decade.

Although carbon and silicon are both in Group 4, their oxides are very different. Carbon dioxide, CO_2, is a gas, whereas the related compound silicon dioxide (silica), SiO_2, found in the Earth's rocks, is a solid. **Chemical Ideas 5.2** explains why this is so.

Activity A9.2 will help you understand the difference between covalent molecules, such as CO_2, and covalent network structures, such as SiO_2.

Coping with carbon

The oceans do a good job in absorbing much of the carbon dioxide we emit, but even so the concentration of carbon dioxide in the atmosphere continues to rise steeply.

Figure 41 is a dramatic illustration of how steep this rise has been. Over the last million years, most of the long-term variations can be explained by changes in the Earth's orbit around the Sun. Over much of the last 1000 years, the cycles of change can be explained by changes in solar radiation and major volcanic eruptions. Usually, a rise in carbon dioxide concentration follows a temperature rise because of positive feedback – as the temperature rises carbon dioxide escapes from the oceans and causes further warming.

The steep rise in carbon dioxide concentration in the second half of the twentieth century is unprecedented – and does not follow a rise in global temperature. Clearly, what is happening now is different to what has happened in the past.

In 1750, before the Industrial Revolution had really started pumping carbon dioxide into the atmosphere, the atmospheric concentration of carbon dioxide was about 280 ppm.

By the early 1990s, the concentration was 353 ppm – and by 2007 it was 383 ppm. Unless drastic action is taken, it is likely to have doubled from its pre-Industrial Revolution value by the second half of this century – within the lifetime of most people reading this.

Climate change models predict that such a doubling would result in temperature rises of between 2 °C and 4.5 °C (with a best estimate of about 3 °C). Although 3 °C may not sound much, it will be enough to have a dramatic effect on the global climate.

The link between the atmospheric concentration of carbon dioxide and global warming, discussed in **Section A7**, is now well established and supported by scientific evidence. While there is still uncertainty about exactly how these changes will affect climate change, the 2007 IPCC report concluded that serious changes to the global climate are very likely and issued stern warnings of:

- reduction in snow cover and thawing of permafrost regions in northern latitudes
- melting of sea ice in polar regions, particularly in the Arctic, with a consequent rise in sea levels

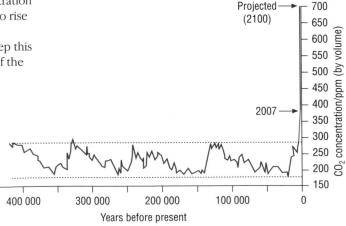

▶ **Figure 41** The concentration of carbon dioxide in the atmosphere over the last 400 000 years (from ice core measurements).

▲ **Figure 42** Melting ice in the Arctic.

- increase in extreme weather such as heat waves, heavy rainfall in northern latitudes, but less rain in tropical regions
- more intense tropical typhoons and hurricanes.

Some of these changes are already apparent. They will affect not only human populations but animal and plant species too.

But how do we go about coping with the carbon dioxide problem? The Kyoto agreement was a start but it did not include several key countries, and there was no provision to support developing countries. The success of the Montreal Protocol in limiting emissions of ozone-depleting chemicals shows what can be achieved by international cooperation. The problems of climate change, however, are much more complex and have a direct effect on the economies of nations – the price to pay for inaction may be very high.

The obvious solution is to reduce carbon dioxide emissions by burning less fossil fuel. This is easier said than done. Options include using less energy, improving energy efficiency, using more renewable and low-carbon energy sources or capturing the carbon dioxide from burning fossil fuels and burying it deep in the ocean. Each has advantages and limitations and it is unlikely that any single approach will provide the entire solution.

Scientists are optimistic that a workable way forward can be found. It will need a global agreement that includes a range of strategies. What is clear is that this will involve changes in our behaviour and the way we use fossil fuels.

Reducing the rate of increase of carbon dioxide in the atmosphere is a major challenge for the world at the beginning of the twenty-first century. In **Activity A9.3** you can investigate some of the possible approaches to solving the problem.

A10 *Summary*

In this module you have looked at two problems in atmospheric chemistry – the destruction of the ozone layer and an increase in the greenhouse effect leading to global warming. These are problems that affect the whole world.

Fundamental to the module is an understanding of the interaction of electromagnetic radiation with matter.

Studying the rates of chemical reactions is vital for chemists working to understand the many reactions taking place in the atmosphere. This led you to consider the factors that affect the rate of chemical reaction – in particular, the effects of concentration and temperature.

Ozone depletion is a *stratospheric problem*. At its centre is the idea of the absorption of radiation by gas molecules. Dangerous effects arise from the increased transmission of ultraviolet light through the atmosphere as ozone molecules are destroyed by radical reactions in the stratosphere.

The mechanisms of many chemical processes in the stratosphere require an understanding of radicals and their reactions. Chlorine radicals produced from CFCs act as catalysts for the depletion of ozone. This led you to study homogeneous catalysis in more detail and to study the role of catalysts in providing an alternative route for a reaction with a lower activation enthalpy.

The greenhouse effect is a *tropospheric phenomenon*. Again it arises from the absorption of radiation by gas molecules. You focused on the role of carbon dioxide and the need to monitor its concentration. This led you to the key idea of chemical equilibrium and its importance in determining the point of balance in some chemical processes. A comparison of the properties of carbon dioxide and silicon dioxide led to a discussion of covalent molecules and covalent network structures.

You can use **Activity A10.1** to check your understanding of the electromagnetic spectrum and the interaction of radiation with matter.

Activity A10.2 will help you to check that you understand the ideas in this module.

POLYMER REVOLUTION

Why a module on 'Polymer Revolution'?

This module has three themes. First, it tells the story of the polymer revolution – the many beneficial changes that polymers have brought to our lives. Many of the discoveries that led to important advances were made by chance; the experiments designed by chemists did not always go according to plan or give the expected result! This is the second theme of the module. The polymers which are produced in the largest quantities are addition polymers, made from alkene monomers. The historical development of addition polymers is the third theme. Most of their development has taken place since the end of the Second World War.

The module provides essential information about polymers and the process of addition polymerisation. You will also learn the necessary terms required to understand and discuss the properties and uses of polymers. Through this study you are introduced to alkenes and their reactions, *E/Z* isomerism and ideas about intermolecular forces. A consideration of some newer polymers will lead to a look at the reactions of alcohols and how infrared spectroscopy can be used to help to determine the structure of organic compounds.

Central to the module is the relationship between the properties of a substance (how it behaves) and its structure (how the atoms are arranged) and bonding (the way the atoms are held together). This relationship helps to explain the properties of polymers.

Overview of chemical principles

In your earlier studies you will probably have come across ideas about:
- polymerisation
- how the structure of a substance determines its properties.

In this module you will learn more about:
- the chemistry of organic molecules (also in **Elements from the Sea**, **Developing Fuels** and **The Atmosphere**)
- isomerism (also in **Developing Fuels**)
- reaction mechanisms (also in **Elements from the Sea**, **Developing Fuels** and **The Atmosphere**)
- electronegativity and bond polarity (also in **Elements of Life** and **Elements from the Sea**).

You will also learn new ideas about:
- addition polymers and polymerisation
- reactions of alkenes
- *E/Z* isomerism (*cis-trans*)
- reactions of alcohols
- intermolecular forces
- the relationship between the properties of a substance and its structure and bonding
- infrared spectroscopy.

PR
POLYMER REVOLUTION

PR1 *The start of the revolution*

Polymers are produced in profusion by nature – in plants and animals and in our bodies. Synthetic polymers are so much part of our lives, both in terms of materials and culture, that it is difficult to believe that their development began as recently as the 1940s. Indeed, polymers have only been in widespread use since the 1950s.

▲ **Figure 1** Plastics touch our lives in many ways.

In the late nineteenth century, plastics were produced by modifying natural polymers. Celluloid, for example, was produced by reacting cellulose (from plants) with nitric acid. The first plastic to be made in significant quantities from manufactured chemicals was *Bakelite* (made from phenol and methanal). Bakelite is still used to make electrical fittings such as sockets and plugs. Although it was first made in 1872, 'by accident', it was not until 1910 that the process was patented and Bakelite was manufactured.

The important polymers which we now know – such as nylon, Terylene, poly(ethene) (polythene), poly(vinyl chloride) (PVC), polystyrene and polyurethanes – were developed many years later between the 1930s and the 1960s. They were not manufactured in significant quantities for the general public until the 1950s. Their growth is part of the story of the development of the chemical industry after the end of the Second World War in 1945, when crude oil began to take over from coal as the main raw material for organic compounds.

Chance has played an important part in the polymer revolution. In this module you will meet a

▲ **Figure 2** **a** A radio made from Bakelite.
b A celluloid film reel.

number of polymers, such as poly(ethene) and poly(tetrafluoroethene) (Teflon), that have been discovered 'by accident'. These chance discoveries resulted from experiments that gave unexpected results, or from errors made by the chemists involved.

The key feature is that the researchers recognised the importance of what they had observed and went on to investigate further. In the A2 module **Designer Polymers** you will meet some polymers that have been specifically *designed* by chemists for particular purposes.

WHAT IS A POLYMER?

A **polymer** molecule is a long molecule made up from lots of small molecules called **monomers**.

If all the monomer molecules are the same, and they are represented by the letter A, then an A–A polymer is formed:

$$-- A + A + A + A -- \rightarrow --\ --A--A--A--A--\ --$$

Poly(ethene) and PVC are examples of A–A polymers.

If two different monomers are used, and they are represented by the letters A and B, then an A–B polymer is formed, in which A and B monomers alternate along the chain:

$$-- A + B + A + B -- \rightarrow --\ --A--B--A--B--\ --$$

Nylon-6,6 and polyesters are examples of this type of A–B polymer. Polymers formed from more than one type of monomer are called copolymers.

Writing out the long chain in a polymer molecule is very time consuming – we need a shorthand version. This is how this is done for PVC, poly(vinyl chloride), now called poly(chloroethene):

poly(chloroethene)

The same basic unit is continually repeated in the chain, so the chain can be abbreviated to:

So this module has three themes: the story of the many beneficial changes brought to our lives by polymers; the importance of chance discoveries in scientific research; and the story of the historical development of addition polymers.

It turns out that many of the polymers which have been found by accident are A–A polymers, known as **addition polymers**. Those that have been designed are mostly A–B polymers, known as **condensation polymers**.

In **Activity PR1.1** your teacher will demonstrate how a polymer is made. In **Activity PR1.2** you will explore some of the polymers that we make use of in our everyday lives and begin to see the link between their properties and their uses.

Chemical Ideas 5.6 will help your understanding of these ideas.

where n is a very large number which can vary from a few hundred to many thousand. The part of the molecule in brackets is called the **repeating unit**.

Plastics, elastomers and fibres

Plastics are polymers which at some stage during their manufacture are capable of flowing when heated, under pressure if necessary. Some plastics repeatedly soften on heating and harden on cooling. These are called thermosoftening plastics, or *thermoplastics* for short. Poly(ethene) and PVC are common thermoplastics.

Other plastics undergo permanent chemical changes when they are manufactured. Bonds, called 'cross-links', are formed between polymer 'chains'. These materials are called thermosetting plastics, or **thermosets**. They do not change shape on heating, but char (blacken) when the temperature becomes high enough. *Bakelite* was the first synthetic thermoset to be manufactured.

Some polymers are soft and springy; they can be deformed and then go back to their original shape. This is because there are fewer 'cross-links' between the polymer chains than in thermosets. Such polymers are called **elastomers**. Natural rubber is an elastomer.

Stronger thermoplastics, which do not deform easily, are just what you want for making clothing materials: some can be made into strong, thin threads which can then be woven together. These thermoplastics, such as nylon, are called **fibres**.

Poly(propene) is on the edge of the thermoplastic–fibre boundary. It can be used as a thermoplastic like poly(ethene), but it can also be made into a fibre for use in carpets.

PR2 *The polythene story*
An accidental discovery

Imperial Chemical Industries (ICI) was formed in 1926 by the amalgamation of a number of smaller chemical companies. The prime aim of the merger was to form a strong competitor to the huge German chemical company IG Farben.

To understand this section you will need to know about a compound called ethene and some of its relatives in the series of compounds known as the **alkenes**. **Chemical Ideas 12.2** contains the information you will need.

In **Activity PR2.1** you investigate the reaction of alkenes with bromine. **Activity PR2.2** will help you to understand how these addition reactions of alkenes take place.

In 1930 Eric Fawcett, who was working for ICI, got the go-ahead to carry out research at high pressures and temperatures aimed at producing new dyestuffs. His results were disappointing and his project was eventually abandoned.

His team then moved into the field of high-pressure gas reactions and was joined by Reginald Gibson. On Friday 24th March 1933, Gibson and Fawcett carried out a reaction between ethene and benzaldehyde using a pressure of about 2000 atmospheres. They were hoping to make the two chemicals add together to produce a ketone:

benzaldehyde   +   ethene

Their apparatus leaked and at one point they had to add extra ethene. They left the mixture to react over the weekend.

They opened the vessel on the following Monday and found a white waxy solid. When they analysed it they found that it had the empirical formula CH_2. They were not always able to obtain the same results from their experiment: sometimes they got the white solid; on other occasions they had less success; and sometimes their mixture exploded, leaving them with just soot.

The work was halted in July 1933 because of the irreproducible and dangerous nature of the reaction.

Learning to control the process

In December 1935 the work was restarted. Fawcett and Gibson found that they could control the heat given out during the reaction if they added cold ethene at the correct rate. This kept the mixture cooler and prevented an explosion. They also found that they could control the reaction rate and relative molecular mass of the solid formed by varying the pressure.

A month later they had made enough of the material to show that it could be melted, moulded and used as an insulator.

Most crucial of all was the identification of the role of oxygen in the process – this was done by Michael Perrin, who took charge of the programme in 1935. When oxygen was not present, the polymerisation did not occur. Too much oxygen caused the reaction to run out of control.

The trick was to add just enough oxygen. The leak in Fawcett and Gibson's original apparatus had accidentally let in a small amount of oxygen. If this had not happened then the discovery of poly(ethene) might not have been made. It was also Perrin who showed that even if benzaldehyde is left out of the reaction mixture, the polymer still forms.

Poly(ethene) is an example of an addition polymer. You can find out more about the formation of polymers like poly(ethene) in the section on addition polymerisation in **Chemical Ideas 12.2**.

Poly(ethene) – or 'polythene' as it is commonly called – is tough and durable, and has excellent electrical insulating properties. Unlike rubber, which had previously been used for insulating cables, poly(ethene) is not adversely affected by weather or water.

It also has almost no tendency to absorb electrical signals. Its first important use was for insulating a telephone cable laid between the UK mainland and the Isle of Wight in 1939. Its unique electrical properties were again essential during the Second World War in the development of radar.

▲ **Figure 3** Poly(ethene) was essential in the development of radar equipment in the Second World War.

The first poly(ethene) washing-up bowls appeared in the shops in 1948 and were soon followed by carrier bags, squeezy bottles and sandwich bags. Sadly, poly(ethene) and some other early polymer materials were over exploited. They were used for all manner of novelties and as cheap but poor substitutes for many natural materials. This gave *plastic* a bad name – the word is often used to describe something which looks cheap and does not last. The reputation still sticks as an undeserved slur on many of today's excellent materials.

▲ **Figure 4** Poly(ethene) bags coming off the production line.

Assignment 1

a Write an equation for the formation of poly(ethene) from ethene.

b Draw the structural formula of part of a poly(ethene) chain. Explain why poly(ethene) can be thought of as a very large alkane.

c Explain why polymerisation of ethene does not occur when there is no oxygen present, but with too much it gets out of control.

A bonus of being big

A polymer molecule is just a very big molecule. This seems obvious now, but the idea wasn't proposed until 1922 and it met with considerable criticism for the rest of that decade. Since the large molecules in a polymer are chemically similar to much smaller ones, it should be possible to predict many of a polymer's properties by looking at the properties of substances which contain the smaller molecules.

Poly(ethene) is the simplest polymer from a chemical point of view, containing only singly bonded carbon and hydrogen atoms. It should behave like an alkane of high relative molecular mass – and in many ways it does. It burns well and tends not to react with acids or alkalis.

Like all polymer materials, poly(ethene) is a mixture of similar molecules rather than a pure compound, because different numbers of monomers join together in the chain-building process before polymerisation stops. Therefore, poly(ethene) does not melt sharply – it softens and melts over a temperature range.

However, this happens roughly at the temperature that you would expect for a very large alkane.

In contrast, poly(ethene)'s mechanical properties are completely unlike those of similar but smaller molecules.

At this point you need to understand about the attractions which arise between poly(ethene) molecules. These attractions (which result from a type of *induced dipole force*) arise between *all* molecules and can be revised in **Chemical Ideas 5.3**. Other types of intermolecular forces are also discussed here.

Activity PR2.3 provides an opportunity for you to check your knowledge and understanding of the work you have covered in **Sections PR1** and **PR2** of this module.

PR3 *Towards high density polymers*

The poly(ethene) produced by Fawcett and Gibson was what we today call low density poly(ethene) (LDPE).

Although they had some control over the product of their polymerisation process, the low density poly(ethene) they made was still quite messy at a molecular level. The polymer chains were extensively branched. This makes it impossible for the chains to fit together in an organised way: they coil around randomly taking up a lot of room, and hence lower the density of the material.

This relatively disorganised and open structure also lowers the strength of the poly(ethene). The next major advance in poly(ethene) production came with the development of **high density poly(ethene) (HDPE)**, which resulted from discoveries made by Karl Ziegler.

The German scientist Ziegler was born in Helsa, near Kassel, in 1898. Encouraged to work hard by his father, the young Ziegler set up a chemical laboratory at home where he became so advanced in his chemistry studies that he was allowed to omit the first year of his degree course.

At the age of 23 he became a professor and started on a research path that was to have a great influence on the future development of polymerisation processes.

Ziegler catalysts

Ziegler was studying the catalytic effects of organometallic compounds. These are compounds which contain covalent metal–carbon bonds. Strange

▲ **Figure 5** Karl Ziegler (1898–1973) who, with Giulio Natta, won the Nobel Prize for Chemistry in 1963 for their work on catalysis for polymerisation.

things happened in some of his experiments when he was using an aluminium organometallic compound.

He tracked down the unexpected behaviour to tiny traces of nickel compounds left over as impurities in his apparatus after cleaning. His research group then tried putting as many different transition metal salts as possible with the alkylaluminium compound to see what would happen.

In 1953 Ziegler, with his colleagues Holzkamp and Breil, found that adding titanium compounds led to the easy production of very long chain polymers. Simply passing ethene at atmospheric pressure into a solution of a tiny amount of $TiCl_4$ and $(C_2H_5)_3Al$ in a liquid alkane caused the immediate production of poly(ethene).

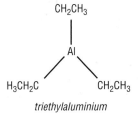

triethylaluminium

Like Fawcett and Gibson's discovery of poly(ethene), Ziegler's achievements began accidentally with an impurity.

The poly(ethene) which Ziegler produced had an average relative molecular mass of 3 000 000 with very

little branching along the polymer chain. The chains could therefore line up and pack more closely than those made in the original high-pressure process. In this form the poly(ethene) is said to be crystalline. This gives the polymer a higher density and greater strength.

High density poly(ethene) is often used to make washing-up bowls, water tanks and piping. It is strong and can easily be moulded into complicated shapes: car petrol tanks, for example, can be made to fit neatly into the spaces under the car, something which was impossible with metal tanks. HDPE is not as easily deformed by heat as LDPE and an early use was as Tupperware food storage containers. The ability to retain shape during heating means that HDPE articles can be heat-sterilised, making HDPE an important material for hospital equipment such as buckets and bed-pans.

Assignment 2

Some data for low density and high density poly(ethene) are given in the table below. Use this information to answer the questions that follow.

	Density/ $g\,cm^{-3}$	Tensile strength/ MPa	Elongation at fracture (%)
LDPE	0.92	15	600
HDPE	0.96	29	350

a Will either of the polymers sink in water?
b Why is the tensile strength lower for LDPE than for HDPE?
c Use the data to explain the different uses of LDPE and HDPE.

Ziegler patented high density poly(ethene). As a result he became a multimillionaire, and on his seventieth birthday he gave $10 million to support further research at the Max Planck Institute where he worked.

Natta and stereoregular polymerisation

Giulio Natta was born in Imperia, Italy, in 1903. He studied chemical engineering at the Milan Polytechnic Institute, and after working at various universities he returned to Milan in 1938 to become Professor of Industrial Chemistry.

Natta was convinced that alkylaluminium catalysts were the key to making stereoregular polymers – polymers with a regular structure.

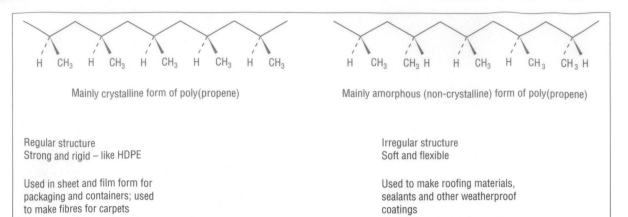

Mainly crystalline form of poly(propene)

Regular structure
Strong and rigid – like HDPE

Used in sheet and film form for
packaging and containers; used
to make fibres for carpets

Mainly amorphous (non-crystalline) form of poly(propene)

Irregular structure
Soft and flexible

Used to make roofing materials,
sealants and other weatherproof
coatings

▲ **Figure 6** Crystalline and amorphous forms of poly(propene).

▲ **Figure 7** Giulio Natta (1903–79).

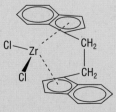

▲ **Figure 8** Poly(propene) of a mainly crystalline form
can be drawn into fibres and used to make carpets.

THE STORY CONTINUES INTO THE NEW MILLENNIUM

Now, a new generation of catalysts for addition
polymerisation is being introduced from the
research laboratory into industry. They are called
the metalocenes. Chemists have compared their
structure to a sandwich. An example is shown
below.

Here the 'filling' of the sandwich is zirconium (but
it could be another transition metal, such as
titanium). The two slices of 'bread' are flat organic
molecules with arene ring systems.

Metallocenes are even more specific than the
original Ziegler–Natta catalysts and they allow

In March 1954 he used Ziegler's catalyst to
polymerise propene. His reaction mixture contained
two forms of poly(propene) – a crystalline form and an
amorphous (non-crystalline) form. He was able to
separate them.

Figure 6 tells you more about these two forms of
poly(propene).

Natta went on to develop new catalysts known as
Ziegler–Natta catalysts. These have allowed chemists
to tailor-make specialist polymers with precise
properties. You will learn more about crystalline
polymers at A2.

chemists to control the polymer's molecular mass as well as its structure.

Poly(ethene) produced using a metallocene can be used as thin films with very interesting properties. They are even more impermeable to air and moisture than the older polymers. They are also strong and much more tear-resistant than the polymers formed by the other methods. Thus thin films of them are used to protect materials that are susceptible to air and moisture, such as food.

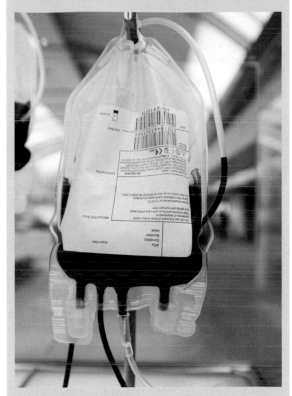

▲ **Figure 9** Polymers made using metallocenes are not easily broken down by γ-rays and so are an ideal material for medical packaging, such as blood bags, which are often sterilised using γ-radiation.

Assignment 3

A very low density poly(ethene), VLDPE, is produced using metallocenes. The chains are essentially linear with a large number of short side chains. The side chains are made by copolymerising ethene with an alkene such as but-1-ene or hex-1-ene.
a   Draw the structures of but-1-ene and hex-1-ene.
b   Draw the structure of the repeating unit formed when:
    i   ethene and but-1-ene react
    ii   ethene and hex-1-ene react.
c   Suggest which of the two copolymers in part **b** has the lower density. Explain your choice.

Assignment 4

Some 'cross-linking' can be introduced into HDPE when it is made into tubing during the moulding process. The polymer is called PEX.

PEX can withstand higher temperatures than HDPE, has better chemical resistance and does not flow so easily. Tubing made from PEX can be bent to an angle of 90° easily without affecting its properties. PEX is increasingly replacing copper for both gas and water plumbing applications.
a   What evidence is there that PEX is an elastomer rather than a thermoset?
b   Suggest reasons why PEX is replacing copper for plumbing purposes.

PR4   *Dissolving polymers*

If soiled laundry from a hospital is mishandled then there is a risk of infection. The risk can be avoided by making the laundry bags out of a dissolving plastic. The dirty linen is safely contained until the bag is placed in the wash – then the bag dissolves and the washing is let out.

The dissolving plastic which is used is **poly(ethenol)**.

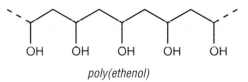

poly(ethenol)

To understand why poly(ethenol) dissolves in water, you need to know about **hydrogen bonding**, a particularly strong type of intermolecular bonding.

You can read about hydrogen bonding in **Chemical Ideas 5.4**.

Activities PR4.1 and **PR4.2** will help to illustrate the ideas about intermolecular forces and hydrogen bonding.

▲ **Figure 10** This hospital laundry bag has a section made of dissolving plastic, which dissolves in the washing machine to release the dirty washing.

Poly(ethenol) molecules contain hydroxyl groups like those in alcohols. You can read about the physical properties and some chemical reactions of alcohols in **Chemical Ideas 13.2**.

You can investigate some reactions of alcohols in **Activity PR4.3** and check your understanding of alcohol reactions in **Activity PR4.4**.

You might think that poly(ethenol) could be made by polymerising ethenol (CH_2=CH–OH), but this compound does not exist. However, it can be made from another polymer, poly(ethenyl ethanoate), by the process illustrated in Figure 11.

The plastic's solubility depends on the percentage of OH groups present. Table 1 shows how the two are related. Different solubilities give the plastic different uses.

Table 1 Solubility of poly(ethenol).

% of OH groups	Solubility in water
100–99	insoluble
99–97	soluble in hot water
97–90	soluble in warm water
below 90	soluble in cold water

Molecules with the structure R–O–$COCH_3$ are examples of *esters* (you will study ester groups in more detail in the A2 module **What's in a Medicine?**). Can you see why the process is called an ester exchange?

You can study the behaviour of poly(ethenol) in **Activity PR4.5** and explore the properties of 'slime' in **Activity PR4.6**.

Assignment 5

a What type of intermolecular bonding will there be between chains of poly(ethenol)?

b Explain why nearly pure poly(ethenol) is insoluble even in hot water.

c Suggest how the intermolecular bonding in this polymer would be affected if the ester groups were still present?

d Explain the effect on solubility of increasing the number of ester groups in the polymer.

e Hospital laundry bags are made from the form of the polymer which is soluble in only hot water. Suggest why this form is chosen.

▲ **Figure 12** An assortment of liquid tablets used for clothes washing.

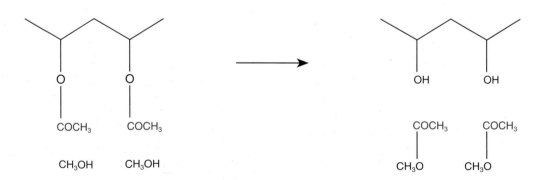

Methanol reacts with poly(ethenyl ethanoate)

Some of the ester groups on the side chains of the polymer are converted to –OH groups and a new ester is formed from methanol

▲ **Figure 11** Formation of poly(ethenol) by ester exchange.

On the shelves of every supermarket are liquid tablets used for washing clothes. These contain measured quantities of liquid detergent for use in domestic washing machines. The sealed film holding the detergent is made from a 'dissolving' polymer.

Assignment 6

Liquitabs are 100% soluble between 30–95 °C.

The instructions on boxes of liquitabs containing detergent include the following statements:
- put the liquitab in the drum before the laundry
- avoid contact with eyes; detergent is an irritant
- handle with dry hands.
a Explain the reasons for each of these instructions.
b Use Table 1 to suggest what percentage of OH groups would be suitable for a poly(ethenol) polymer used for making liquitabs. Explain your choice.

PR5 *Polymers that outdo nature*

Poly(ethene)- and poly(propene)-based plastics and fibres are often seen as replacements for natural materials. Today there are many addition polymers that can be used to do jobs where nature fails. In this section we shall look at two of these important discoveries.

The Teflon man

Roy Plunkett worked in the research laboratories of DuPont, a large chemical company in the US. On 6th April 1938 he wanted to use some tetrafluoroethene, a gas stored in a cylinder. However, the cylinder appeared empty. He decided to open the cylinder and discovered a white waxy solid. The gas had polymerised to form poly(tetrafluoroethene), PTFE, which is now marketed as 'Teflon'.

$$n \, \underset{\substack{\text{F} \\ | \\ \text{F}}}{\text{C}} = \underset{\substack{\text{F} \\ | \\ \text{F}}}{\text{C} \longrightarrow} \left(\begin{array}{cc} \text{F} & \text{F} \\ | & | \\ \text{C} - \text{C} \\ | & | \\ \text{F} & \text{F} \end{array} \right)_n$$

tetrafluoroethene *poly(tetrafluoroethene)*

On examining the polymer, Plunkett discovered its now well-known anti-stick properties. But Teflon is also highly resistant to chemical attack and is a very good electrical insulator.

▲ **Figure 13** PTFE is commonly used as a coating for non-stick cookware.

Water out but not in

Poly(tetrafluoroethene) is a hydrophobic (water-hating) material. In 1969, Bob Gore discovered that the polymer could be stretched to form a porous material that would allow water vapour but not water liquid to pass through the minute holes.

He developed a material – Gore-tex – which uses this property. A layer of porous PTFE film and a layer of an oil-hating polymer act as the filling of a sandwich between an outer fabric and the inner lining (Figure 14). The oil-hating polymer allows the water vapour through, but prevents the natural oils from the skin and from cosmetics blocking the pores in the PTFE, thus preserving its waterproofing properties.

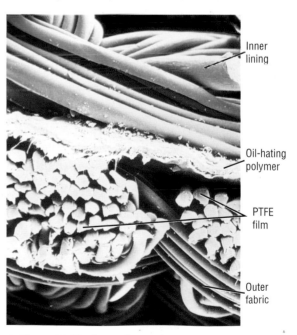

Inner lining

Oil-hating polymer

PTFE film

Outer fabric

▲ **Figure 14** The layers that make up a Gore-tex membrane.

▲ **Figure 15** Gore-tex is ideal clothing for bad weather conditions. It allows the water vapour from the body to evaporate.

ETFE

Chemists have designed a copolymer of ethene and tetrafluoroethene called ETFE. The discovery of this material made it possible to build the two giant transparent domes, called *biomes*, to house plants from all over the world at the Eden Project in Cornwall. Inflated panels of ETFE were used because of their

▲ **Figure 16** The Eden Project – a biome covered with ETFE cushions.

unique properties: high transparency, very low density, high resistance to intense radiation for a long time period, shatterproofing and stain resistance. Glass had already been ruled out because it could be damaged by radiation and could shatter. In addition, a large glass dome would be too heavy to be supported adequately.

Neoprene

Neoprene was first synthesised commercially by DuPont in the early 1930s. It was used as a rubber substitute in motor car components. Today it is used in making wetsuits. It is also the preferred material for making cases for MP3 players, personal digital assistants (PDAs) and mobile phones. As an elastomer, it has many properties that make it far superior to natural rubber in numerous applications. It is non-porous, very tough and resistant to light, heat and chemical attack. The essential constituent of Neoprene is a polymer made by polymerising chloroprene to produce poly(chloroprene).

▲ **Figure 17** Neoprene in use today.   **a** A surfer's wetsuit is waterproof, tough and helps to insulate the body from heat loss.   **b** A case for an MP3 player absorbs shocks.

CH$_2$=C—CH=CH$_2$
    |
   Cl

or

CH$_2$=C—CH=CH$_2$

Cl

chloroprene

The properties of Neoprene can be modified by copolymerising chloroprene with 2,3-dichloro-1,3-butadiene. It is also treated with sulfur to introduce some necessary cross-linking.

You can read about copolymerisation in **Chemical Ideas 5.6**.

Assignment 7

a Give the systematic chemical name of chloroprene.

b Draw part of a polymer chain made from one molecule of chloroprene and one molecule of 2,3-dichloro-1,3-butadiene.

c Suggest why cross-linking is introduced into Neoprene.

The properties of polymers such as Neoprene are affected by the polymerisation process. This is because of the way different groups can be arranged around a carbon–carbon double bond. Before you go any further you will need to have some understanding of an area of chemistry which may be new to you. *E*/*Z* isomerism is a phenomenon which arises in alkenes and a number of other types of compound.

You can read about *E*/*Z* isomerism in **Chemical Ideas 3.4**.

Natural rubber or gutta-percha?

Natural rubber is an addition polymer of 2-methyl-1,3-butadiene; it is usually called *isoprene*. The polymer is formed naturally through a complex series of reactions and is called poly(isoprene). In the rubber tree, *Hevea brasiliensis*, the poly(isoprene) chains are mainly formed with the *Z* (or *cis*) arrangement of groups around the double bond.

2n

isoprene
2-methyl-1,3-butadiene

natural rubber
(Z)-poly(2-methyl-1,3-butadiene)

A Malaysian tree, *Palaquium gutta*, also produces a poly(isoprene) polymer. However this material, *gutta-percha*, is hard and non-elastic. This is because the arrangement of the groups around the double bonds is *E* (or *trans*) rather than *Z*. This material has been used for coating golf balls and early underwater cables.

gutta-percha
(E)-poly(2-methyl-1,3-butadiene)

Activity PR5.1 will help you to make models of *E*/*Z* isomers and understand the differences in some of their properties.

Chemists can identify the repeating units present in polymers using infrared spectroscopy. You can find out how infrared spectra can be interpreted by reading **Chemical Ideas 6.4**.

Activity PR5.2 will then test your skills in interpreting the infrared spectra of some polymer samples.

PR6 *Summary*

In this module you have seen how our ideas about addition polymers have developed. As a result of this development, chemists have learned how to build up simple molecules into very long polymer chains. The polymers they have produced are new materials with unique sets of properties not possessed by any natural substances.

You began by reading about the accidental discovery of poly(ethene), which led you to a study of alkenes and their reactions. You saw how the properties of poly(ethene) depend on its structure and the intermolecular forces between the polymer chains. The instantaneous dipole–induced dipole forces that are present in all substances are important here. For other addition polymers with polar groups on the chain (such as PVC), permanent dipole–permanent dipole forces are also present and the attractive forces between the chains are stronger.

You went on to find out about the discovery of Ziegler–Natta catalysts and the use of metallocenes to control the polymerisation process. These developments have allowed chemists to control the structure of the polymer formed – and hence control its properties.

Dissolving polymers have hydroxyl (OH) groups on the polymer chain. To explain why they dissolve in water you needed to find out about hydrogen bonding, a particularly strong type of intermolecular attraction.

You were introduced to an important group of synthetic elastomers and learned about *E/Z* isomerism in alkenes and how it is important in deciding the properties of such polymers. Finally, you learnt how to use IR spectroscopy to identify the functional groups in polymer samples.

One of the themes of this module was the accidental nature of many of the important discoveries in polymer chemistry. They came about because the scientists involved recognised that they had observed something unusual and interesting, and went on to investigate further.

Activity PR6 allows you to check that you understand the ideas contained in this module.

INDEX

Note: **CS** = Chemical Storylines AS;
CI = Chemical Ideas.

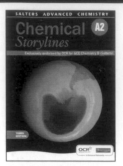